Computer systems in work design
— the ETHICS method

Computer systems in work design — the ETHICS method

Effective Technical and Human Implementation of Computer Systems

ENID MUMFORD and MARY WEIR

A work design exercise book for individuals and groups

A HALSTED PRESS BOOK

JOHN WILEY & SONS
New York—Toronto

English language edition, except USA and Canada published by
Associated Business Press, an imprint of Associated Business Programmes Ltd.,
Ludgate House,
107-111 Fleet Street
London EC4A 2AB

Published in USA and Canada by
Halsted Press, A division of John Wiley & Sons Inc.,
New York

First published 1979.

Library of Congress Cataloging in Publication Data

Mumford, Enid.
Computer systems in work design—the ETHICS METHOD.

"A Halsted Press book".

1. Work design—Data processing. I. Weir, Mary, Joint author. II Title.
T60.8.M85 658.31'42 78-32068
ISBN 0 470 26656 2

Typeset by Photo-graphics, Honiton, Devon, and Molineaux (K. Studios) Ltd., Coulsdon, Surrey.
Printed and bound in Great Britain by
A. Wheaton & Co., Ltd., Exeter

Contents

	Foreword		vii
	Preface		ix
1	**Introduction**	*Background to the ETHICS method; outline of this book*	1
2	**Job satisfaction**	*A major objective for the systems design process*	5
3	**The ETHICS method**	*For diagnosing needs and designing work systems*	26
4	**Exercise 1**	*Babycare Ltd, Food Packaging Department*	45
5	**Exercise 2**	*Mail-It Ltd, customer accounting system*	129
6	**Exercise 3**	*International Bank, Foreign Exchange Department*	199
7	**Conclusions**	*The ETHICS method and its practical applications*	287
	Appendix A	*Notes for using the exercises with groups of students*	296
	Appendix B	*Glossary of computer terms used in the case studies*	302
	References		310
	Index		313

Foreword

by David Firnberg
Director
The National Computing Centre

The computer revolution is gathering pace, and new systems will be affecting more and more people at their work and in their daily lives.

Those of us who are close to computing are aware of the potential benefits and of the problems that may arise.

If we wish to ensure that the potential is exploited fully for the benefit of people, we must make sure that the requirements of the people who will work with the new systems are properly considered.

Enid Mumford's and Mary Weir's research is particularly important, as it has resulted in some techniques and methods that will enable the requirements of people to be considered during the design and implementation of new systems. Too often the systems designer concentrates on the technical aims of the system, but then at implementation the manager has to adjust the work to suit the system. The result can be that the economic gains from the system are not realised, and that people find that they are not suited to the new tasks. As the authors say there is a 'bad fit' between the employees' expectations and the new job requirements, with resulting lack of job satisfaction.

Now that computing equipment is becoming cheaper and more flexible, the technical constraints are being reduced. The methods described in this book show how both the technical and human requirements can be analysed, and systems designed which people both can and will be pleased to use.

I therefore welcome this book and commend it to all those involved in developing new systems. I hope it will be widely read, and that the methods described will be applied to the benefit of all computer users.

Preface

This book has been prepared to give you, the reader, some practical experience in using a particular approach to work design. This approach takes both technical and human factors into account in the design process and attempts to arrive at a solution that caters well for both efficiency and job-satisfaction needs. In doing this it avoids a tendency, common in much job design, to regard technology as fixed and immutable, with the result that the design processes concentrate solely on improving the social system and accept any constraints stemming from the technical system.

1 Introduction

Background to the ETHICS method

The authors of this book have for many years been interested in the sociological aspects of introducing new systems of work.

Our earlier studies were concerned with the implementation of computer systems, and how this could be achieved with the minimum disruption and discomfort for those involved. Our particular interest centred on the employees whose jobs were being altered by the computer system, and the kind of education, training and communication policy which were needed to help them adapt to the new work system. We were also concerned with the changes in salary structure and working conditions which had to be negotiated, as well as the nature of the mechanisms for consultation which were set up.

As a result of these studies we learned a great deal about the conditions and policies which could facilitate major change such as the introduction of a new kind of technology into an organisation. Many of our findings have been written up or will shortly be available.

However, one of our observations particularly intrigued us and has led us on to our present research work. In some of the organisations we studied, the computer system was being installed in several locations, or a single computer system was serving several different locations in the same way, and yet the jobs of employees associated with these systems differed quite significantly in the different locations. Some of the jobs had much more interest and variety even though they were dealing with input or output from the same computer system.

The variations in jobs stemmed largely from the different philosophies of the managers of the user departments, and their perceptions of what constituted a job which was efficient for the organisation and, if the manager considered this important, satisfying for employees. In most cases, jobs were designed by the local user manager almost as an afterthought when the technical system was being implemented, and needed human assistance to complete the tasks which the computer could not do. But by that time all kinds of constraint had been unwittingly built into the technical computer system, and these considerably curtailed his freedom to organise the work in a way which would create interesting jobs. The other difficulty was that the user manager had little or no social-science training and therefore did not use the body of research knowledge which has been built up on the design of jobs.

These observations led us to realise that our efforts to assist in the smooth implementation of computer systems might be of little value or even harmful, if the computer system itself had been poorly designed in the first place, with little thought given to the jobs which would be created as a result. Our interest then moved to the area of work design, since we felt that only by considering the jobs associated with computer systems *at the design stage* could the needs and expectations of employees be adequately catered for. We are now moving on from computer systems to other forms of technology and other kinds of systems design.

The research which we are currently undertaking is based on two propositions:

1. That many kinds of technology (and this is particularly true of computer technology) are sufficiently flexible to allow for the design of systems which take into account the needs of employees for satisfying work. Therefore both the technical and human parts of a work system should be designed with this objective in view.

2. That even in situations where the technical system has been designed and implemented, it is still possible to re-design jobs in ways which will make them more satisfying.

Whenever possible, we feel that it is desirable for new work systems to be designed using the philosophy set out in our first proposition. But where this is not feasible, then there are techniques available such as job rotation, job enlargement and job enrichment which can be used to improve jobs without altering the technical system. We group these techniques in the second proposition under the general heading of systems adjustment.

In our role as teachers, we have used our research findings to put forward our conviction that computer and other technical systems can and should be designed in such a way as to improve the quality of working life for employees. But, perhaps not surprisingly, we have met with some resistance from systems designers, engineers and managers who feel that we are asking the impossible when we suggest that they should take into account the needs of people at work, at the very time when they are wrestling with the immense problems of designing a new technical system. Technology, they say, is complex enough without bringing the even more unpredictable human being into it as well.

Even where there is agreement with our ideas and a willingness to try and cater for human needs, there are the practical difficulties of actually doing this, especially when the training of most systems designers, engineers and managers includes very little, if any, social science.

To help systems designers, managers and other interested groups, and for our own use, we have developed what we call the ETHICS method, an acronym for the Effective Technical and Human Implementation of Computer Systems. This method is still being developed through action-research projects, but is sufficiently far advanced to be of practical assistance to organisations introducing new systems of work or improving old ones, and who wish to try and use this opportunity to improve the job satisfaction of their employees.

Briefly, the ETHICS method consists of a set of steps which must be taken in the design and implementation of a new work system. At each stage, technical *and* human needs are taken into account, so that the system is designed specifically to meet both technical and human objectives at one and the same time. We have already used this method in many organisations, and are undertaking further research to develop and refine it.

Outline of this book

In order to explain the ETHICS method, we have over the years developed a series of case-study exercises as a teaching aid. These exercises, which are based on real situations, enable students to try out the method for themselves. We have found that students both enjoy doing them, and acquire a clear understanding of what work design really means.

The exercises have been conducted with many different groups of people of widely differing backgrounds and experience, and yet the majority have been able to cope with the ideas quite easily and have found that this way of learning can be relatively painless. The graduates and managers at the Manchester Business School have been our main participants, as well as groups of systems designers, computer specialists, personnel managers, prison officers and administrators, and even a group of Spanish businessmen.

This book sets out a detailed account of the ETHICS method in Chapter 2. Here the theoretical basis of the method is described and its various stages drawn together. Chapter 3 explains how the case-study exercises should be used. Chapters 4, 5 and 6 present three of these exercises and their suggested solutions, although for reasons of space these case studies cover only the diagnosis and design stages of the ETHICS method in detail. Chapter 7 summarises the ideas which we are attempting to impart, and points the way for further improvement in systems design.

2 Job satisfaction

A major objective for the systems design process

'What merit is there in an over developed technology which isolates the whole man from the work process, reducing him to a cunning hand, a load bearing back, or a magnifying eye, and then finally excluding him altogether from the process unless he is one of the experts who designs and assembles or programs the automatic machine? What meaning has a man's life as a worker if he ends up as a cheap servo-mechanism, trained solely to report defects or correct failures in a mechanism otherwise superior to him? If the first step in mechanisation five thousand years ago was to reduce the worker to a docile and obedient drudge, the final stage automation promises today is to create a self-sufficient mechanical electronic complex that has no need even for such servile non-entities.'
Lewis Mumford
The Pentagon of Power[27]

Work design — the need for a new approach

This book is an attempt to get everyone concerned with the design of new systems of work to consider human as well as technical and business factors when setting design objectives.

In other words, to take what social scientists call a 'socio-technical' approach to the systems-design process. A socio-technical approach has been defined by Davis as one which recognises organisations as purposive entities with a wide variety of goals which, in order to survive, have to interact successfully with their surrounding social and technological environments.[9] Such an approach also perceives that individuals and groups have their own needs and values and that these must be met if they, in turn, are willingly to satisfy the needs of the organisation. Technology is an important variable in this network of relationships which both affects the ability of the individual or group to meet its own personal needs and the ability of the organisation to interact successfully with its business environment. A socio-technical approach to work design is one which recognises the interaction of technology and people and produces work systems which are both technically efficient and have social characteristics which lead to high job satisfaction.

At present much of the work-design process does not take this socio-technical view. Davis criticises the approach of conventional systems designers, saying:

> 'No clear objectives concerning roles for men as men are visible, although objectives are clearly defined for men as machines. When man is considered only as a link in a system, design rules do not exist for allocating appropriate tasks to man. Nor are there design rules for designing task configurations to make complete and meaningful jobs . . . If we turn to how technology is translated into requirements for job designs, we see widespread acceptance of the notion of the technological "imperative", put forth by most engineers and managers. That a substantial part of the technical design of production systems involves social system design is little understood or appreciated.'[10]

Computer systems design, in particular, although it has been with us for only a comparatively short period of time, appears to have become structured and formalised and to embody a problem-solving philosophy which is accepted uncritically by systems designers. This philosophy sees the design of computer systems as a technical process directed at solving problems which are defined in technical terms. Existing manual systems are seen as too inaccurate, too slow or too labour intensive and their replacement by a computer-based solution is regarded as a means for increasing efficiency and reducing costs.

Computer systems design, at least at the data-processing level, usually takes the form of breaking down the manual system into its component parts, identifying which of these are essential to the problem-solving function the department is there to perform, and reassembling these in a form appropriate to the logic of computer usage. The new system will usually include a more rigorous set of controls than the old to ensure that it works with the required speed and accuracy. Work systems based on other forms of technology, such as assembly-lines, also take this technically orientated approach and the human being is expected to conform to the demands of the machine.

It is interesting to speculate how this particular design philosophy has arisen and become so widely accepted. It seems to have its roots in the analytical methods of work study and in the early scientific-management attempts to reduce human error by breaking work down into small bits and tightening work controls; logical at a time when skilled labour was in short supply. It also seems to owe part of its approach to an outdated engineering philosophy of designing machines and assuming that the human being will adapt to their requirements, whatever these may be. More recently this systems-design philosophy has been reinforced by the training given to systems analysts and engineers which appears to ignore the fact that unless a work system is completely automated, it is a *man-machine* system. That is a system in which men and machines work together in a production function; the man cannot operate at a high level of efficiency without the machine, the machine cannot operate at all without the man. Such a system is unlikely to function effectively if this mutual dependency goes unrecognised and only the machine part of the system is consciously designed.

Social scientists are continually surprised that the very intelligent people associated with the design of new systems of work and with the training of systems designers and engineers continue to regard the current design approach as tenable and legitimate. For example, there are, as yet, few attempts to introduce into computer training courses subjects which would seem to be of the utmost relevance to the EDP discipline.[1] These include general systems theory, the socio-technical approach to systems design, ergonomics and work design.[22] All of these subjects now have a considerable literature associated with them.

Yet perhaps one must not be too critical of existing practice. For many years past, engineering systems have been designed with little attention paid to the job-satisfaction needs of the worker operating the production process. These systems were introduced with the implicit assumption that the human being would and must adapt to the demands of technology. Managers and academics have tended to accept this approach uncritically and, while a great deal of research has been directed at identifying the impact of different kinds of technology on work organisation, until recently little attention has been paid to establishing how a technology can be manipulated so as better to provide for human needs.[3, 19, 31, 34]

Today, industry is greatly worried by the strength of the conflict between management and workers and is increasingly becoming interested in the possibilities for reducing work alienation and increasing job satisfaction through policies of job enrichment and worker participation in operational decision-taking. The Scandinavians have taken these ideas further than any other country and have developed work systems which not only provide stimulating and challenging tasks but also give work groups responsibility for most of the decisions which affect the way their work is planned and carried out.[14] In Britain we have had a number of experiments in job enrichment and a new form of production organisation called 'group technology' is being tried out by a number of firms.[29] One version of this, developed by Edwards, involves the design of a production system from the start as an integrated man-machine or socio-technical system.[11]

If a more humanistic approach to work is not soon adopted by engineers and systems designers, it can be predicted that many existing and new technology-based work systems will become increasingly unacceptable to workers and their trades-union officials; and also to managers, who will not welcome work systems with built-in industrial-relations problems. A work system that is designed to achieve objectives defined solely in technical terms is likely to have unpredictable human consequences. The reason for this is that technical decisions taken at an early stage of the design process will impose constraints on the organisation of the human part of the system. Because the human part of the system is either not included in the design process, or not considered until a very late stage, the human consequences of these decisions may not be recognised until the system is implemented. This can lead to the technical system influencing the human system in a way which was never envisaged by the systems designers.

Engineers and computer systems designers do not yet appear to realise that with their present approach they are designing only partial systems and that to do their job effectively they must be competent systems designers in both the technical and human areas. Already trades unions in Europe are recognising that the way work systems are designed is of the utmost importance to the job satisfaction of their members and a number are making the quality of working life a factor in their bargaining processes.

If the argument is accepted that the traditional work-design philosophy is untenable in the long term because it offends modern humanistic values and because it is becoming the cause of serious industrial-relations problems, then a new approach needs to be considered — one which not only attempts to avoid human ill effects but has the positive aim of encouraging the setting and achievement of human objectives as an integral part of the work-design process. Such an approach implies that those associated with the introduction of new work systems — engineers, computer systems designers, managers, trades-union officials and clerical or production workers — are prepared to take up a value position which accepts that technology can and must be used to achieve human as well as technical objectives. Any work systems which do not meet this double criteria will be rejected as unacceptable in human terms.

In an attempt to show that such a humanistic approach is viable, the authors and their research colleagues have developed a method for designing work systems in which equal weight is given to human and technical needs and human as well as technical objectives are set. This method will be described in the following chapter.

In order that technology can be exploited to serve human ends, particularly those of the individual in work, a great deal needs to be known about the design opportunities and limitations of different forms of technology. In addition, there needs to be a willingness on the part of managers, engineers and systems designers to accept the achievement of human objectives as an integral part of the work-design process, and a recognition on the part of trades-union officers that the exclusion of human objectives from work design can lead to work procedures which incorporate characteristics which are undesirable from a human point of view. Sackman who has written a great deal on the subject of computer systems design, tells us 'humanistic automation means that computers serve human ends and that automation is not an end in itself. Computers are there to elevate man's intellect and increase his control over his environment.'[30]

The research of the authors has been directed at the development of a set of principles which assists the systematic and integrated design of both the technical *and* human parts of any system so that both technical efficiency *and* job satisfaction are increased. We believe that our approach will provide organisations with a number of benefits. It will enable them to introduce technical change with less stress and upheaval than is usually the case and, in doing this, it may reduce the financial costs of innovation. Most important, it will help ensure that technology is used in a positive way to improve the quality of working life for all employees.

A definition of job satisfaction

If we are to make the improvement of job satisfaction a major goal of the work-design process, then it is important to know precisely what we mean by the term. A universally acceptable definition of job satisfaction is extremely hard to find and we have therefore developed our own. We define job satisfaction as the attainment of a good 'fit' between what the employee is seeking from his work — his job needs, expectations and aspirations — and what he is required to do in his job — the organisational job requirements which mould his experience. (See Figure 2.1.)

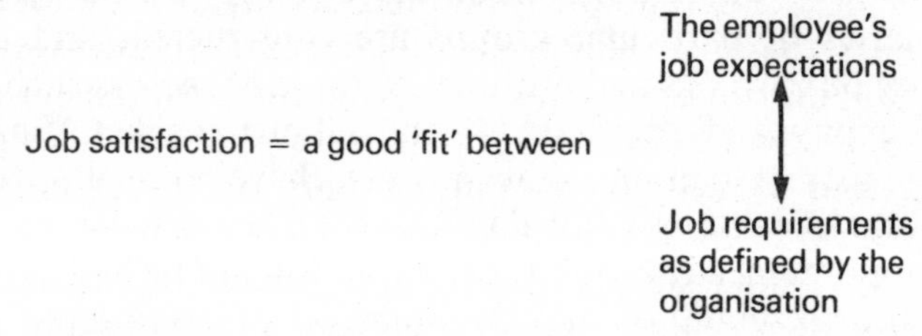

Figure 2.1

If we are to design new work systems or alter existing ones in such a way as to improve job satisfaction, then we need to identify some significant factors which can be measured in order to check how good or bad the 'fit' is between an employee's job expectations and the requirements of his job both before and after the new system is implemented or the existing one modified. We need, in effect, to have a viable theory or framework for measuring job satisfaction. Unfortunately no coherent theory of job satisfaction exists and one has had to be developed to meet the needs of this research.

Examination of the literature on job satisfaction shows that it is split into a number of different schools of thought, each with its own particular focus. First, there is what we will call the psychological-needs school. Those psychologists such as Maslow, Herzberg, Likert, etc., who see the development of motivation as the central factor in job satisfaction and concentrate their attention on stimuli which are believed to lead to motivation — for example, the needs of individuals for achievement, recognition, responsibility, status.[16, 17, 20] A second school devotes its attention to leadership as a factor in job satisfaction. Psychologists such as Blake and Mouton see the behaviour of supervision as an important influence on employee attitudes and they therefore direct their observations at leadership style and the response of subordinates to this.[2]

A third school represented by such experts in wage-payment systems as Lupton, Gowler, Legge and Bowey approach job satisfaction from a quite different angle and examine the effort-reward bargain as an important variable.[4, 5, 13, 18] This leads to a consideration of how the wages and salaries of particular groups are constructed, and the influence on earnings and attitudes to these of factors such as overtime pay and the state of the labour market. Some psychologists maintain that people have a subjective perception of what is a fair day's work. They believe that if this is not obtained then job satisfaction will not be high.

The fourth school of thought approaches job satisfaction from an entirely different angle and sees management ideology and values as an important influence. Writers such as Crozier and Gouldner identify different value systems in organisations.[8, 12] For example, Gouldner categorises certain forms of management behaviour as 'punishment-centred', 'representative' and 'mock' bureaucracy.

Punishment-centred bureaucracy is the type of management behaviour which responds to deviations from rules and procedures with punishments. Representative bureaucracy is the kind of management practice which today would be called 'democratic'. Here rules and procedures are jointly developed by management and workers to meet a group of shared and mutually agreed objectives. Mock bureaucracy is when an organisation has rules and procedures but neither management nor workers identify with these or accept them as legitimate. Consequently they are generally ignored. Although a discussion of values as such does not appear often in the job-satisfaction literature, it is clear that the ethics and moral philosophy of a company, together with the kind of legislation that management formulates and employee perceptions of the legitimacy of this, must have an influence on job satisfaction.

Fifth and last, there are behavioural scientists who say that the factors described above are extrinsic to the tasks an employee is required to carry out, and therefore a less important factor in job satisfaction than the work itself and the way it is structured. This group concentrates on the content of work and on job design factors. Amongst their numbers in Britain are Cherns, Cooper, Wilson and the members of the Tavistock Institute and Work Research Unit. The European pioneers of this approach are to be found in Norway and include Herbst, Thorsrud and Gulowsen.[6, 7, 13, 15, 32, 33] American pioneers include Davis, Hall and Lawler.[9, 14]

All the different schools of thought described above have made important contributions to the theory of job satisfaction although they may not have consciously set out to do so. For example, Lupton, Gowler, Legge and Bowey would regard their work as a contribution to control theory with earnings as a key factor, and not just as a contribution to job-satisfaction theory. Similarly the Crozier, Gouldner school are examining different manifestations of bureaucracy and only indirectly relate these to concepts of job satisfaction. Nevertheless, a coherent theory of job satisfaction must embrace all these ideas.

A job-satisfaction model was developed some years ago by the Computer and Work Design Research Unit at the Manchester Business School. This is based on the job-satisfaction theories set out above. These have been integrated into a conceptual framework derived from the work of Talcott Parsons, which is described in detail elsewhere.[23, 28] Parsons sees organisations and individuals as constantly making choices when they are presented with different situations. The choices are the following:

1. Between seeking immediate gratification or deferring this until a future date.

2. Between seeking to further interests private to oneself or interests shared with others.

3. Between deciding to accept generalised standards in the interests of conformity and control or to seek for an acceptance of individual differences and a unique approach which may be a response to emotion rather than intellect.
4. Between evaluating people and things because of what they are — their attributes — or because of what they do — their achievements.
5. Between choosing to react to a period or a situation in a widely differing manner according to the perceived requirements of the situation or reacting to a situation in a limited and specific way. For example, an individual may seek a variety of satisfactions from work or require satisfaction only on the earnings variable — a fair day's wage for a fair day's work.

These five choice areas* of Talcott Parsons are helpful when considering job satisfaction for two reasons. First, Parsons uses them at three levels; that of personality, social system and cultural system. This means that they can be used to describe the orientation of an organisation, a group or an individual. Second, as Parsons claims, they are very comprehensive categories, and they appear to cover all the factors that researchers have considered to be associated with job satisfaction. Unfortunately, as currently phrased and defined they are intellectually cumbersome and need some redefinition and simplification before they can be used operationally to study job satisfaction.

A first attempt at this simplification led to the following set of categories. † These were related to what Parsons calls an individual or group's need dispositions — what the individual or group seeks from the work situation.

---

* Given the collective title of 'pattern variables' by Parsons and individually named as follows: Affectivity — affective neutrality; Self-orientation — collectivity orientation; Universalism — particularism; Ascription — achievement; Specificity — diffuseness.

† These categories were originally defined in terms of the 'fit' between company and individual needs. In this article we are considering individual needs alone and so only this part of the definition is given here.

1. *The individual's job requirements.* What the individual seeks from his employer if his most important needs are to be met. Some of these needs will be urgent and immediate, others will be deferred. That is, the individual will hope to achieve a number in the short term and the others in the long term.

2. *An orientation towards self-interest.* The extent to which an individual wishes to pursue his own interests in the work situation, and is unwilling to forgo these if they come into conflict with his employer's interests.

3. *An orientation towards individuality.* The extent to which an individual wishes to behave in a unique and individual way (to express his own individuality, 'do his own thing'), seeks a work situation which allows him to do this and is unwilling to accept uniform policies, methods and standards imposed by his employer.

4. *The desire for an employer to recognise personal qualities of character and personality.* The extent to which an individual wishes to be recognised for what he is, as opposed to what he does.

5. *Opportunities for work flexibility.* The degree of work flexibility which an individual requires to match his skills, knowledge and personality. Work flexibility is related to technology and requires an absence of technical constraints which control the number and kind of tasks an individual performs and the manner in which he performs these.

It can be seen that we are now starting to approach a coherent framework for examining job satisfaction, although even these new definitions are somewhat vague, especially number 1 which appears to embrace all the rest. A clearer set of categories which brings together Parson's analytical theory and the ideas of the different schools of thought on job satisfaction is set out in Figure 2.2. Job satisfaction is again defined as the achievement of a good fit between job needs and expectations and job experience.

If an employee's needs in these five areas are met then we

can hypothesise that he has high job satisfaction. The 'fit' between employee needs and expectations and employee job experience may be good on all five variables or it may be good on some and poor on others. If the 'fit' is not entirely good then the question 'Why is this?' must be asked and answered by the manager, engineer or systems designer. The cause may lie in poor personnel policies or it may lie in an unsatisfactory task structure. If the latter is the case then remedial measures should be possible through the design of the new work system.

On the employee's side, he may see his needs as not being satisfactorily catered for because he has unrealistic expectations of what he may reasonably require of his employer. On the employer's side, the environment in which the firm is operating at a particular moment in time may prevent him meeting some of his employee's needs. For example, the economic climate may prevent him providing what his employees regard as an equitable effort-reward bargain; the technology his firm uses may make it impossible for him to meet employee needs for a task structure which permits interesting and varied work, or increased competition in the firm's product market may make it essential to tighten quality standards and controls and increase output. All of these things can inhibit the ability of the employer to satisfy his employees and can result in a bad 'fit' between what employees want and what they experience in work.

If there is a bad 'fit' on any of these variables then the 'fit' can be improved by the opportunities for change presented by the new or redesigned work system. The psychological, efficiency and social-value 'fits' can be improved through alterations in personnel policies as well as through job design. The knowledge and task-structure 'fits' can be improved primarily through a socio-technical approach to work design which creates forms of work organisation and job structure related to people's needs. These then are the areas which are measured before the introduction of the new work system in order to establish if there is a good 'fit' between job requirements and employee job needs and expectations. They are measured again once the new system has been implemented and has settled down to establish if the use of

	The employee's job needs	*The employee's job experience*
	The employee:	*A good 'fit' exists when he:*
The KNOWLEDGE 'fit'	Wishes the skills and knowledge he brings with him to be used and developed.	Believes that his skills and knowledge are being used and developed to the extent he wishes.
The PSYCHOLOGICAL 'fit'	Seeks to further interests private to himself e.g. secure: achievement, recognition, responsibility, advancement, status.	Believes that his personal interests are being successfully catered for.
The EFFICIENCY 'fit'	Seeks a personal equitable effort-reward bargain, and controls, including supervisory ones, which he perceives as acceptable. Seeks efficient support services such as information, supervisory help.	Believes that financial rewards are fair and other control systems acceptable. Believes that he receives the support services he requires to do a competent job.
The TASK-STRUCTURE 'fit'	Seeks a set of tasks which meets his requirements for task differentiation, e.g. which incorporate variety, interest, targets, feedback, task identity and autonomy.	Has a set of tasks and duties which meet his needs for task differentiation.
The ETHICAL (social-value) 'fit'	Seeks to work for an employer whose values do not contravene his own.	Believes that the philosophy and values of his employer do not contravene his own values.

Figure 2.2

socio-technical design principles, together with any necessary alteration in personnel policies, has improved the aspects of work which were previously unsatisfactory and leading to poor job satisfaction. Let us examine each area separately:

- The *knowledge* 'fit'.
- The *psychological* 'fit'.
- The *efficiency* 'fit'.
- The *task-structure* 'fit'.
- The *ethical* (social-value) 'fit'.

The knowledge 'fit'

The employee is prepared to use his skills and knowledge to further his employer's interests. In return he expects that his employer will not under-utilise these assets; but will assist him to increase them, so that he feels competent to handle more complex work problems.

There appear to be considerable differences in the extent to which people recognise their own skill and knowledge potential and wish this to be fully utilised, and the manager, engineer or systems designer must diagnose the needs of each situation in which he intervenes. For example, research into the attitudes of clerks has shown that many older women clerks are looking for an easy life in work and do not want to be mentally stretched. In contrast, another clerical group, bank clerks, who entered employment with good education qualifications, complained that their jobs were too easy and that they were not able to use fully the skills and knowledge which they possessed. A socio-technical approach has been used in the design of a new computer system in this bank and a task structure has been associated with the computer system which permits skills and knowledge to be better utilised and developed. Our third exercise tackles the design problem at this bank.

There will be a good 'fit' on this variable when the employee believes that his skills are being adequately used and that he is being assisted to develop them and become increasingly competent.

There will be a poor 'fit' if the employee believes that his skills and knowledge are being under-utilised, and his opportunities for personal development restricted.

The psychological 'fit'

Herzberg has provided a great deal of important research data which shows that if an employee is to be motivated by and satisfied with work, then his employer must meet his needs for recognition, achievement, responsibility, status, advancement and work interest (the Herzberg motivators).[16] The theory behind this research of Herzberg is that we all have powerful psychological needs, many of which we seek to gratify within the work situation. If the employing organisation can ascertain and meet these needs then it will develop motivated and satisfied employees.

Psychological needs are influenced by a variety of personal factors including sex, family background, education and class. They tend to vary over an individual's life cycle so that the needs of a person starting his career are likely to be different from those of a person nearing retirement age. The studies of motivation undertaken by Herzberg suggest that certain psychological needs are common to a majority of people.[16] Herzberg and another psychologist, Maslow, also believe that there is a hierarchy of needs and that as certain basic needs become satisfied so they become less urgent and are replaced by others which are seen as having greater urgency.[16,20]

The redesign of work which is a result of a new or modified technical and social system can help the achievement of a better 'fit' in this area. Job enlargement and enrichment can provide greater opportunities for feelings of achievement, responsibility and status. The redesign of work cannot meet all these psychological needs, however, and attention will also have to be given to personnel policies related to career development and to other aspects of the work situation.

There will be a good 'fit' on this variable when the employee believes that his personal aspirations for recognition, achievement and other psychological need factors are being adequately met within the work situation.

There will be a poor 'fit' if the employee has psychological needs related to work which the work situation cannot provide for.

The efficiency 'fit'

The organisation seeks employees who will meet its productivity and quality standards and who will accept its administrative procedures and controls. The wage-payment system is one element in the control process, for through this a financial agreement is struck with the employee that buys his conformity. In return the employee requires a fair financial reward and for these procedures and controls to be arranged in such a way that he retains a degree of personal influence over his activities.

Here we have a contractual relationship based on the organisation's need to produce goods or services and the individual's willingness to meet these providing:

- The effort-reward bargain is seen as fair and his economic needs are met.
- Work controls are seen as reasonable — neither too rigid nor too loose.
- Supervisory controls are acceptable.

The effort-reward bargain is the amount that a firm is prepared to pay to get the skill and competence it requires, set against the evaluation of individuals of how much their skills are worth, and their expectations of what they are likely to receive. This contractual area has traditionally been seen by management as the most important and the one with greatest influence on employer-employee relationships. Yet studies of many white- and some blue-collar groups have shown that in certain circumstances employees will place financial rewards low down on their list of needs.[26] But if there is dissatisfaction with pay structures and awards, then the introduction of a new work system will provide the opportunity for reformulating these. Also, because new systems change job requirements it is of the utmost importance for management to ensure that wages and salaries are appropriate for the new task structure.

Work controls are our second factor and we find that firms vary greatly in the controls which they use. Some favour tightly structured rules and procedures which they believe reduce the margin of misunderstanding and error. Others leave their staff wide limits within which to set their own targets and monitor their own performance. McGregor, in his analysis of Theory X and Theory Y management styles suggests that Theory Y, with its emphasis on autonomy and self-control is more acceptable and effective than Theory X, but other evidence suggests that people adjust to the kinds of control which are in operation and it is possible that this is not an important factor in job satisfaction.[21] The critical factor may be the relevance of selected controls to the needs of a particular work group or work situation. If a group is used to self-imposed, flexible controls and approves of these then it is important that the new work system does not tighten these. If, in contrast, a group has tight external controls and does not like these, then the systems designers must give some thought to how the work system can be formulated so as to permit a greater degree of self-control by the work groups.

In addition to an effective and accepted control system, efficiency is assisted by a set of support services which help the individual to work in a well organised way with all the necessary back-up facilities which he requires. These will include information, materials, specialist knowledge and supervisory help. Employees who do not receive the support services which they regard as essential to the efficient performance of their job are likely to become frustrated and dissatisfied.

There will be a good 'fit' on this variable when employees achieve the rewards, controls and support services which they require.

There will be a bad 'fit' if the employee believes that he is unfairly paid, undesirably constrained by procedures and controls, or unable to work efficiently because of poor support services.

The task-structure 'fit'

The task-structure 'fit' is the organisation of work activities in such a way that, from a negative point of view, the employee is not required to undertake anything that he regards as too onerous, too demanding, too dull, or too simple; and, from a positive point of view, provides him with a set of challenging tasks and decisions. The 'fit' on this variable will be a good one if the level and kind of work provided by the employer meets the employee needs for stimulus.

This element of job satisfaction is strongly influenced by technology, for many of the jobs a firm requires its employees to perform are directly related to the technical processes it uses to make its product. A particular structure of tasks will lead to simple, routine work, another to work which is complex and challenging. The organisation, in the past, has sought employees who would adapt unquestioningly to its technology and contingent task structure. There was little understanding that the employee might work more effectively in a work situation where he was provided with a level of work variety and challenge which met his needs. Technology has for some time been recognised as exerting a powerful effect on behaviour and attitudes at shop-floor level but its direct influence on the jobs of specialists and managers has been less potent until the advent of the computer.

When a person is allocated a particular work role he is instructed by his supervisor that certain tasks and responsibilities accompany it. Some of these are prescribed — that is, they must be done, but others will be discretionary and the individual has a degree of choice over whether and how he carries them out. The nature of these tasks and responsibilities is clearly an important element in job satisfaction although one that until recently has only received limited attention. Cooper, a psychologist, who has studied job content and design suggests that work can be analysed in terms of:

- The number of skills that need to be used.
- The number and nature of targets that have to be met and the feedback mechanisms that tell the individual when these targets have been achieved.
- The identity of the task as shown by its separation from other tasks by some form of discontinuity or work boundary, and its visibility as an important and meaningful piece of work.
- The degree of autonomy and control that an individual has in the performance of work activities.[7]

Jobs differ greatly in the blend of these four characteristics that they provide. If a particular job mix does not meet an employee's expectations of the kind of work he should be doing then there will be a bad 'fit' between individual needs and job requirements on this task-structure variable and job satisfaction will be reduced.

This is the aspect of job satisfaction most open to improvement through a socio-technical approach to the design of a new work system and most vulnerable to the technically dominated approach. The human part of the system can be designed so as to produce individual jobs which are high in the work characteristics set out above or it can be designed to produce a group structure in which autonomous work groups allocate work activities among their members, set their own targets, take their own decisions, etc.

There will be a good 'fit' on this variable if technology and task structure produce a work situation wherein the employee has the amount of work variety and opportunities for the use of discretion which broadly fit his personal requirements.

There will be a bad 'fit' on this variable if employees who require variety and challenge are expected to work on narrow, specialised jobs. Groups which cannot obtain variety and challenge from their work may obtain this through waging a 'war game' with management on the industrial-relations front — an expensive and dysfunctional process.

The ethical or social-value 'fit'

In work, people are evaluated both for their performance and for their behaviour when relating with others. Some organisations place a great deal of importance on performance and rate personal qualities such as sympathy, trust, integrity, much lower. Other organisations value highly employees who are successful as human beings as well as workers. Organisations seek employees who match their ideologies and cultures. In turn employees seek employers who hold similar sets of values to their own.

Most people wish to be evaluated not only for their performance but also for their qualities as people. For example, their success in making friends, winning respect and inspiring confidence. Some organisations place a great deal of importance on what might be called 'civilised' behaviour in work. Others are more interested in efficiency as an end in itself and want 'efficiency' orientated men and women. Clearly an individual who believes very strongly in the importance of social relationships will not be very happy in a firm which cares only for its production figures and views its employees solely as a means to this end. If he works for such a company his job satisfaction is likely to be low. Similarly, a tough, efficiency-orientated man may become irritated in a situation where he is expected to pay great attention to the feelings and interests of his colleagues and subordinates.

The 'fit' between what an employee wants and what he receives on the ethical/social value variable is an interesting one about which a considerable amount has been written but little related to the subject of job satisfaction. Yet it is likely to be an increasingly important factor in job satisfaction as employees demand better communications and more involvement in decision-taking.

Many of us have strong personal values which have developed throughout our lives and which exert a powerful influence over the way we behave. We believe that there is a 'right' and 'wrong' way of relating with others. Organisations also have values and firms are perceived by the communities in which they operate as 'welfare minded', 'paternalistic', 'ruthless' etc. Values are not easy to change and men and women who have a strong sense of right and wrong may find it difficult to achieve job satisfaction if they have an employer whose values do not coincide with their own on matters which they regard as important. Major differences between management and employee values can lead to internal strife, disruption and what the firm sees as 'undesirable work attitudes'.

The ethical/social value 'fit' is likely to be more concerned with the planning and implementation than the operation of a new work system. It will be important for the employer to meet the employee's needs for communication, consultation and participation in the design of the system. Failure to do this may lead to distrust and apprehension and a feeling on the part of the employees that they are not being treated as they should be. Work systems which reduce an employee's decision taking responsibilities should be regarded as undesirable and efforts made to develop systems which allow the individual to take decisions and provide him with better information by which to do this. There will be a good 'fit' on this variable when the organisation is able to meet those employee values which concern communication, consultation, participation and other aspects of human relationships.

There will be a bad 'fit' if the organisation has employees who place great importance on human relationships and an ethical approach to others, yet are asked to work in an impersonal environment where performance is rated above everything else.

Similarly there will be a bad 'fit' if employees are not human-relations conscious while their employer is.

This job-satisfaction framework provides us with a useful tool which can be used both to measure the overall job satisfaction of a particular group of employees or the 'goodness of fit' between employee job expectations and the demands made of them by their jobs through the way in which these have been structured. In any redesign process there needs to be a good understanding of why certain 'fit' areas are successful; in other words why there is a good match between what the employee is seeking from the work situation on this variable, and what he is receiving. Where job-satisfaction 'fits' are poor then the reason for this must be ascertained and the improvement of this 'fit' made a design objective.

3 The ETHICS method

Our method has the following systematic steps:

- *Diagnosis* of the needs of the social system focusing on long-term job-satisfaction needs. This diagnosis will be used as a basis for setting objectives, developing strategies and for socio-technical systems design.
- Systems *design*.
- Systems *monitoring*.
- Post change *evaluation* of the effectiveness of the systems-design approach.

Diagnosis is based on the analytical approach which we have already described. Using the job-satisfaction model set out in the last section, information is obtained (on our knowledge, psychological, efficiency, task and ethical variables) on the 'fit' between what employees are seeking from the organisation which employs them and what they need to receive if their wants are to be met. If this 'fit' is found to be a good one in the pre-change situation, then it is likely that job satisfaction will be already high, and there should be an absence of serious management-employee conflict. This may assist the acceptance of change as there will be fewer attempts by either side to use the introduction of a new work system to further their own interests in a conflict situation. Against this, a situation where there is a very good job-satisfaction 'fit' may have little internal dynamic for change and the problem for the systems designer who is upgrading technology by, for example, introducing a new machine or computer system, will be to design and present the change in such a way that recipients are assured that they will gain further satisfaction from it amd improve on an already good 'fit'. If the 'fit' is poor, then providing the systems designer is aware of the reasons why this is the case, he can design the new work system to as to achieve a 'human' improvement.

The use of questionnaires as a means for collecting diagnostic data

Information on the 'fit' between employee job needs and organisational job requirements is obtained through questionnaires. Employees are given a questionnaire which they fill in themselves and are told that the data collected in this way will be used by those responsible for the design of a new work system to improve job satisfaction, with the design processes acting as one of the means for doing this.* Questionnaire data are analysed by computer and graphs printed out for each department or section to be affected by the proposed system change showing on which of the five job-satisfaction measures there is a good or bad 'fit'. It will be remembered that the 'knowledge' and 'task-structure' fits can be directly improved through the redesign of work. This is also true of some aspects of the other 'fit' areas. But a poor 'fit' on the 'effort-reward' or 'ethical' variables will also require a change in personnel policies and, with the ethical 'fit', a change in organisational philosophy. In order to understand positive or negative attitudes on the 'task structure' variable the pre-change work system is examined by means of job analysis. This provides observational data on the nature of the tasks which are causing dissatisfaction.

These data are next subjected to democratic examination. Employees who filled in the questionnaires are brought together in small groups, and problem areas which have been identified by means of statistical analysis are discussed and clarified with them. The objective of these meetings is to

* In many firms where this approach has been used the questionnaire data act as one of the first stages of a participative design approach in which groups of shop-floor workers or clerks design their own new work systems.

identify the reasons for dissatisfaction with particular aspects of work and to formulate improvements which can realistically be aimed at through the design of the new work system. It is stressed that wherever possible employees will themselves be encouraged to play a major part in this design process.

Socio-technical systems design*

In this book we set out the ETHICS method as a training exercise. When we are helping in the design of socio-technical systems we use it to assist systems designers, engineers, managers and clerical and shop-floor employees to think logically through the technical and social alternatives available. Once the diagnostic data on job satisfaction have been obtained and discussed with the employees concerned, we ask the design group to use them as a basis for setting some human objectives directed at increasing job satisfaction. We next ask the technical experts in the design group to specify the technical alternatives available for achieving the technical objectives which they have set, which are related to increasing business efficiency, and assess each one in terms of its ability to assist or prevent the attainment of our human objectives.

The next step is to think out the different ways in which the human part of the work system can be organised and to establish the advantages and disadvantages of each approach in relation to the human objectives. Human alternatives will cover the different ways of structuring groups in the department, the mix of tasks that can be allocated to each

* We are using the term socio-technical in a broad sense here. We mean that the technical parts of a work system are carefully evaluated in terms of their human consequences, and technical choices made both in terms of the contribution of the technology to efficiency and its ability to contribute to job satisfaction through permitting a satisfying social structure and mix of jobs to be associated with it. Flexible technologies are more likely to have this characteristic than deterministic ones.

It should be pointed out that our approach differs in a number of ways from the classical socio-technical design procedures developed by experts such as Davis, Emery, Trist and Thorsrud.

group and the design of individual jobs. Once technical and human alternatives have been set out in this way it is then possible to identify those which both fit together *and* achieve both technical and human objectives. The selected socio-technical solution will normally be those technical and human alternatives which jointly best meet technical and human objectives. Experienced design groups may prefer not to separate technical and human alternatives in this way but to think through a series of alternative socio-technical solutions to their work problem. This approach is quite acceptable but, in our experience, more difficult unless the design group has a very good knowledge of the technology they are concerned with and of work design. Our approach is set out in Figure 3.1.

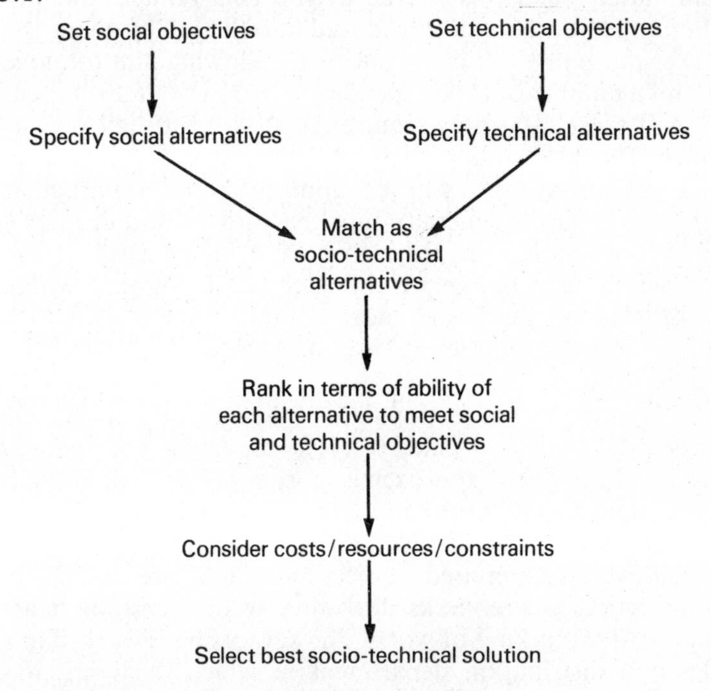

Figure 3.1 Socio-technical systems design

If the technical system is already installed or specified so that it cannot be altered in any way, then the ETHICS method can still be used, although the consideration of technical alternatives will not form a part of the design procedures.* Instead analysis will focus on identifying the impact of the technical system on the jobs of people who interact with it and will set out its social disadvantages and advantages. Technical constraints or imperatives that reduce the number of social alternatives available for consideration will be noted, together with any technical features that can be exploited to design a work structure which provides job satisfaction.

In order to specify human alternatives the group responsible for systems design needs to have a good knowledge of different ways in which work can be organised. Many of these are concerned with improving the employee's 'task variety' and this is true of *job enlargement* and *job rotation*. Here variety is introduced into the job by giving the worker more tasks to carry out or by allowing him to move around a number of tasks, spending a period of time on each. None of these tasks may require much skill for their performance. This kind of approach can be useful in that it reduces work monotony and will be appropriate if a particular employee group is generally satisfied with work and merely wants more variety in order to have a higher level of job satisfaction. It is unlikely to be adequate for groups whose job-satisfaction needs are more complex and related to a desire for opportunities to use skill, meet challenge, and exercise control.

An approach that is popular at the present time is *job enrichment*. This also tends to focus on giving work greater variety, although the approach is more sophisticated. Work is now designed so that the employee is able to use a number of different skills, some of which are quite complex and require judgement to be exercised, choices made and decisions taken. Many offices where clerks deal directly with customers are able to offer this kind of work. The authors have worked in a sales and distribution department in which the work was redesigned when a new computer system was introduced, so

* The technical system can also become given when a systems-design team is permitted to take technical decisions without any consideration of their human consequences, and the design of the human part of the system is then constrained by these decisions.

that all clerks had the same set of responsibilities. These were carrying out a number of routine activities associated with inputting order information into a computer via visual display terminals but also undertaking some more complex activities associated with dealing with customers and handling customer queries; a set of activities which were intricate and demanding and required the use of excellent judgement. In this department the ability to use judgement and take decisions was a product of knowledge, not job grade, so that even the newest clerk could start handling difficult customer problems as soon as he or she felt that they had the knowledge to do so. The clerks in this department also had additional responsibilities in that they were able to requisition their own materials and personally obtain any customer information they required through interrogating the computer via the visual display units. They were able to handle their own problems and organise their own work, and supervisory assistance was only requested when the problem was very difficult indeed.

In this situation work was enriched by putting together two sets of tasks which had previously been handled by different groups and by giving the clerks responsibility for activities, such as material requisition and the obtaining of information, which previously had been carried out by supervision. The role of supervision was now primarily that of long-term planning and co-ordination with other departments in the firm. This kind of job enrichment seems to improve both job satisfaction and efficiency in many situations, providing that employees can be trained to the necessary level of competence. It may be difficult to introduce into departments where there are few experienced employees and where there is a very high level of labour turnover. Also, because this approach does not give employees any responsibility for the development of new methods, it does not lead to any new thinking on better ways of carrying out the work.

It is not easy to incorporate development aspects of work into non-managerial jobs, although intelligent staff are likely to find that an opportunity to develop new ideas and try out new methods is one of the most satisfying aspects of their work. If jobs have this component it is also of great advantage to management as work methods will be constantly reviewed and improved and suggestions will be coming from staff for innovations that assist the prosperity of the business. A way of incorporating development activities into work becomes possible if we stop thinking about individual jobs and turn our attention to autonomous groups or, as the authors prefer to call them, self-managing groups. The focus of our attention is now the group rather than the individual and we are able to consider how we can incorporate more complex responsibilities into the work of a group of non-supervisory staff who see themselves as a team. If we concentrate our attention on the kind of self-managing group that is multi-skilled in the sense that each member is competent to carry out all the operational activities for which the group is responsible, then many things become possible. The group is now co-ordinated and organises its own work so that individual task responsibilities integrate well together and the group works as an efficient team. Such a group is more easily able to initiate and try out new ideas and methods than an individual employee. Similarly, if management has confidence in the ability of its self-managing groups then it can hand over a great many control activities. It can, for example, let the group organise its own work activities and set its own performance targets and monitor these. It can give it responsibility for identifying and correcting its own mistakes. It can give it a budget and allow it to buy its own materials and even organise the selling of its own products to customers. This kind of group may require little supervisory intervention in its activities and management's responsibility will become one of long-term planning and boundary management. By boundary management is meant ensuring that the work of all the self-managing groups in a department is co-ordinated and the work of the department as a whole integrates well with that of other contingent departments.

The self-managing group can be excellent in the right situation and it provides a stimulating work environment in which staff can readily develop their talents. However for it to succeed certain things are necessary. First of all the work of the department into which the self-managing groups are introduced must provide scope for multi-skilled work that provides challenge and responsibility. In many situations the work has been so strictly allocated between departments that no rearrangement of tasks or creation of self-managing groups can make it much more interesting. In this kind of situation the challenging, problem-solving aspect of the work has been separated off and handed over to a specialist group in a separate department. Any real improvement in work interest can now only be achieved if several departments are merged together, thus providing the required work variety. Secondly, the creation of multi-skilled self-managing groups with the competence to control many of their activities requires intelligent, responsible employees together with excellent long-term training. It may take several years to make all members of a group multi-skilled and if the group suffers from a high labour turnover then management will find the training process an expensive one. Thirdly, the creation of self-managing groups has implications for salary levels and grading schemes. Grading can no longer be related to length of service, it must be related to knowledge and skill. Therefore a new employee who is adept at learning all the jobs for which a group is responsible will reach the highest grade quickly and this may be resented by long-serving members of the firm who have worked their way slowly up a hierarchy of grades over many years. But this is a problem associated with a change from one philosophy of work to another and it should not cause permanent difficulty.

Systems monitoring

Once the new system is being implemented it is essential to monitor closely what is happening. Although our diagnostic tool and system-design method can both provide useful starting points for systems change, the implementation process requires careful monitoring to ensure that design strategies are staying in line with human objectives and that objectives set at an early stage of the design process continue to be valid. If the change process deviates from the intended course of giving considerable weight to human needs then mechanisms must be available to bring it back on course.

Systems evaluation

The success of a new work system in human terms is rarely evaluated after implementation is completed. Yet this is essential if planning and systems design are to incorporate a learning process in which the mistakes of the past are avoided in the future.

To assist evaluation our diagnostic tool can be used once again. We can look at the nature of the 'fit' between employee needs and organisational job requirements now that the new system is in. If the post-change 'fit' is better than the pre-change 'fit', then the new work structure has led to an improvement in job satisfaction. If the 'fit' is still unsatisfactory then some remedial action needs to be taken.

We have already used the ETHICS method to assist a number of firms to cater systematically for employees' job-satisfaction needs when introducing new work systems.

In these firms we have been associated with the following:

- Identifying human needs by means of our job-satisfaction diagnostic tool.
- Helping managers and design groups to set human objectives.
- Assisting managers, engineers, computer technologists and other groups to design systems in socio-technical terms to meet technical and human objectives.
- Monitoring the human aspects of systems implementation.
- Evaluating the success of systems in terms of increased job satisfaction.[24, 25]

The case studies — a preliminary explanation

The case studies which follow are all based on the experiences of real firms when introducing new work or computer systems. We were fortunate enough to be able to spend some time with each of these firms and to make a systematic analysis of the nature of the problem each was trying to solve. However, the approach set out in this book and the solutions suggested are not necessarily those that were chosen by the firms. Our solutions are derived from the equal weighting we have given to both human and technical objectives and while one of these firms also had this philosophy, the other two did not.

The case studies cover the diagnosis and socio-technical design stages of the ETHICS method. Though this is only part of the total method, it should be remembered that in the real situations these two stages could take several years to complete, so we are already compressing an enormous amount of work into three short exercises.

We find the exercise works best if undertaken by groups rather than individuals, but this book has been structured so that you can work through the exercise by yourself and in this way gain an understanding of the approach. If you later decide to use it with a group we have found that the best size of group is from five to eight people. Larger or smaller groups are still feasible, but it is better to try and arrange for the exercise to be carried out by two small groups rather than one large group. Guidance for using the exercise with a group of students will be found in Appendix A.

Instructions for doing the case studies

The following description outlines the steps which should be undertaken to complete the diagnosis and socio-technical systems-design stages of the ETHICS method, and is summarised in Figure 3.3.

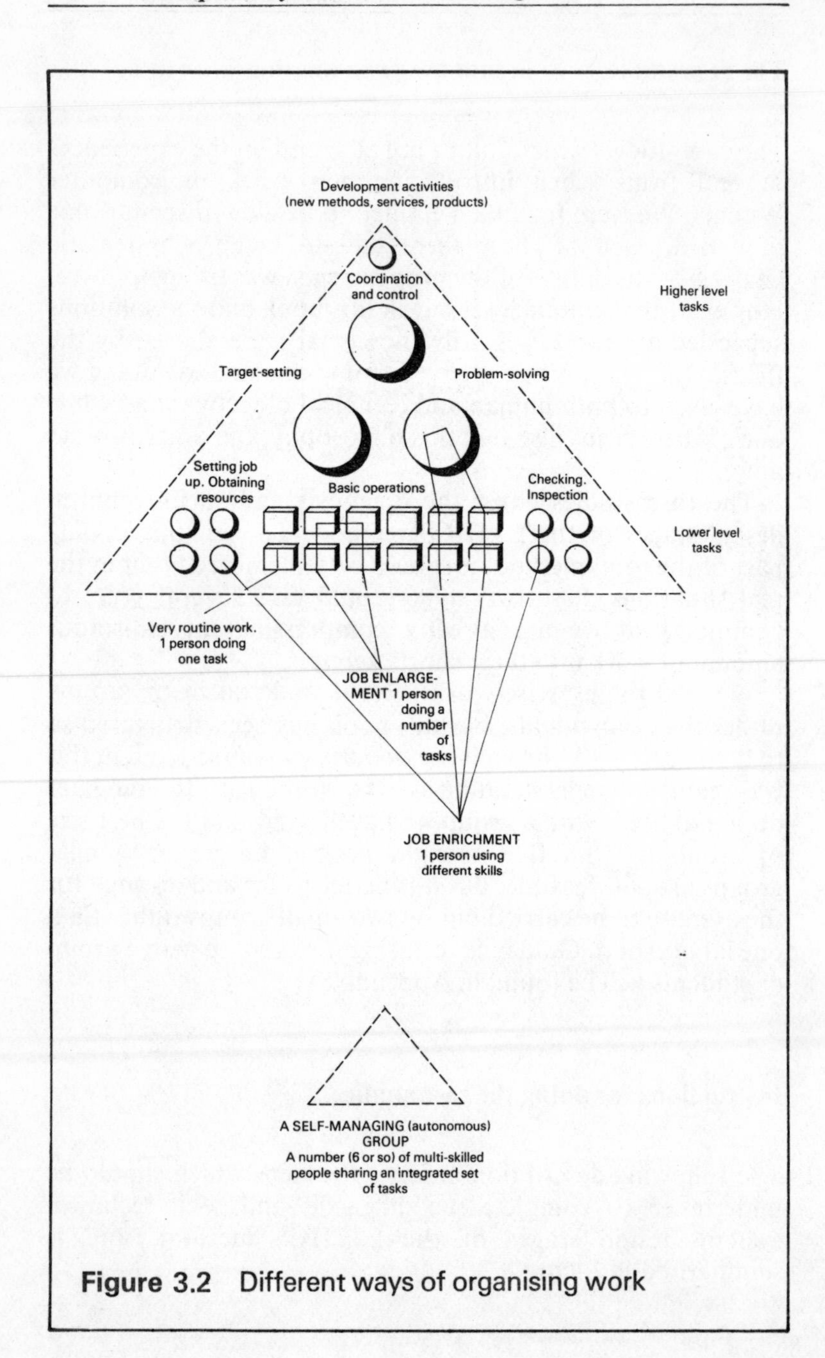

Figure 3.2 Different ways of organising work

Figure 3.3 THE ETHICS METHOD
Diagnosis and Socio-technical Systems Design
The steps that you will be following in the case-study exercises

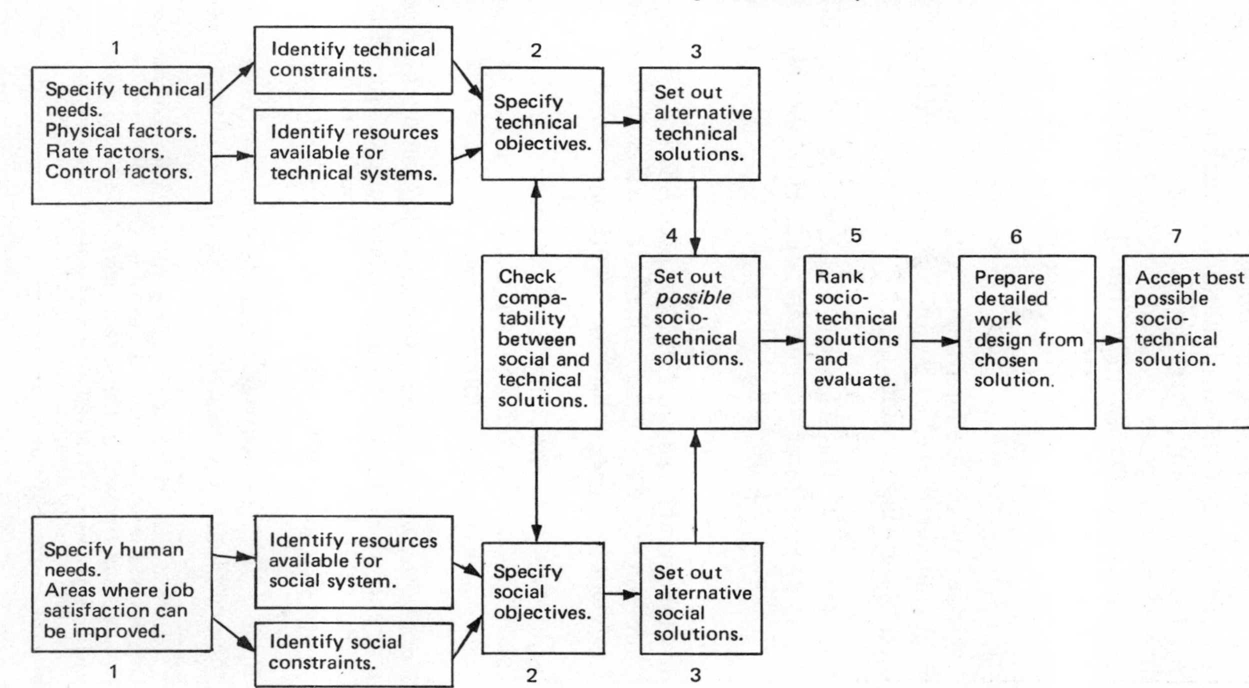

Step 1: diagnosis

The information required for the diagnosis of human needs is collected through the use of questionnaires based on our job-satisfaction framework, which employees complete themselves. The results obtained from these questionnaires are presented for each case study, though in a simplified form to allow the data to be handled easily. The questionnaire results set out what employees like and do not like about their present jobs, and what they would like to have in an ideal job.*

Replies should be analysed in terms of the 'fit' between what employees receive now and what they would ideally like on the following variables:

Knowledge 'fit'. The extent to which they think their skills and knowledge are being adequately used. How they would like their skills and knowledge to be used.

Psychological 'fit'. The extent to which their needs for advancement, recognition, respsonsibility, status and achievement are being met. These are the factors identified by Professor Herzberg as leading to work motivation.

Efficiency 'fit'. The extent to which they perceive wages and salaries to be just and equitable, accept output and quality controls, and believe support services to be effective.

Task-structure 'fit'. The extent to which the work offers the amount of skill and variety, opportunities for decision-taking, autonomy, meaning, and the possibility of achieving clearly defined targets, which they would like to have.

Ethical 'fit'. The extent to which the ethics and philosophy of their employer fit their own. For example, do communication

* In the real-world situations these data would be both comprehensive and complex and would always be discussed in detail with employees before design groups move into the second step of the design process. In the exercises the data are presented in the form of statements with which employees agree or disagree. There are of course many other equally valid ways of designing questionnaires.

and consultation practices conform to that they believe to be right?

Once the questionnaire data have been analysed on all these variables, it is possible to summarise the information and see what employee needs the new system should be designed to meet, so far as is possible. These employee needs should be set alongside the technical requirements which the new system must meet. In our opinion, the technical requirements and social needs should be given equal weight in the design of the system.

Step 2: socio-technical systems design

It is now possible to set human objectives which the new system should achieve, based on the social diagnosis which was completed in Step 1.* You will find that the technical requirements of the system have been specified and are likely to include physical factors, that is the need to produce a departmental product of a particular kind; rate factors, the need to produce the product within certain time limits; and control factors or the need to check that quality and quantity standards are being met. In order that the objectives will be realistic, and capable of achievement without unreasonable difficulties, it is essential first of all to identify the resources which are available for the design, implementation and operation of the new system, as well as any constraints which must be taken into account. For instance, on the social side, one constraint which might be imposed is that the new system will not result in any redundancies of present staff. Such a decision may well have implications for the design of the system, so that it can be operated by staff with the particular levels of skill at present employed, rather than being designed for people with different skill levels, who would have to be recruited from outside the organisation. It would also be

* In all the case-study exercises the technical objectives are given for the following reasons: 1. Many individuals and groups doing the exercise will not have the technical knowledge to set these objectives. 2. The amount of technical information that would need to be given before objectives could be set would be considerable. 3. The emphasis is on designing the social system.

necessary to ensure that sufficient resources were available within the organisation to cope with the retraining programme which would be required for existing staff. These resources and constraints are specified in the case-study exercises.

The diagnosis of job-satisfaction needs together with an identification of required resources and existing constraints precede the setting of design objectives. At this stage these objectives can only be specified in very broad terms, but they provide policy guidelines for future work, and also a check against which the final system can be evaluated.

For example, broad human objectives might be to have an efficient, pleasant office system and to provide job security, job satisfaction and opportunities for employee development.

Step 3: setting out alternative solutions

It is now possible to set out the various solutions which are feasible to achieve the objectives which have been specified in Step 2, while taking account of the resources which are available, and the constraints which must be met.

At this stage, the technical and social solutions are put forward independently, so that the range of solutions considered is as wide as possible. It may be that a broad search at this stage could produce solutions which are outstandingly good for either the technical or social system, so that it will be worth while later trying to modify the other system to accommodate this solution, so long as no overall objectives are sacrificed.

The advantages and disadvantages of each proposed solution are now set out, and it is important that both the technical and the social advantages and disadvantages are listed for each technical and social option. Once four or five reasonable alternative technical and social solutions have been put forward, they should be evaluated separately against the criteria established earlier in Step 2, to see that they are feasible solutions. Each solution that is accepted as feasible can form part of a short-list of technical and social solutions.

Step 4: setting out possible socio-technical solutions

In the previous step all the technical and social solutions were generated independently of one another, so as to ensure a wide range of thinking. Now they must be compared with each other to establish which 'fits' are compatible, so that a short-list of socio-technical solutions can be drawn up. The proposed solutions on the technical list should be compared with those on the social list, and pairs which are in harmony can be noted as possible socio-technical solutions.

Step 5: ranking socio-technical solutions

We now have a list of socio-technical solutions in which the social and technical elements are compatible and which are seen as achieving the objectives established in Step 2 and catering for the job-satisfaction needs in Step 1. It is now necessary to rank these solutions and to select for detailed design the one that fits best. In real-world situations this can be the most difficult task of the whole exercise, since so many variables need to be considered at once. The evaluation which is made can be as simple or sophisticated as time, resources and the importance of the system dictate. A cost-benefit analysis of the various alternatives would normally be undertaken at this stage as well as tests of the hardware being considered. For the purpose of this exercise, however, only the very simplest evaluation is made.

Once the best socio-technical solution has been chosen from the list, the decision must be evaluated against the job-satisfaction needs set out in Step 1, against the technical and social objectives for the system which were put forward in Step 2, and against required resources and existing constraints.

Step 6: preparing a detailed work design

The steps followed so far have set out a system which should meet technical and human needs at one and the same time. However, it is still necessary to give some thought to the design of task structures within the system. Step 6 considers the mix of tasks which will be created for groups and individuals by the chosen socio-technical solution. These tasks should be fitted together so that a work group or individual has a logically integrated pattern of work activities that incorporates the job-design principles set out below:*

Task variety. An attempt must be made to provide an optimum variety of tasks within each job. Too much variety can be inefficient for training and frustrating for the employee. Too little can lead to boredom and fatigue. The optimum level is one which allows the employee to take a rest from a high level of attention or effort while working on another task. Or conversely to allow him to stretch himself after periods of routine activity.

Skill variety. Research suggests that employees derive satisfaction from using a number of different kinds and levels of skill.

Feedback. There should be some means for informing employees quickly when they have achieved their targets. Fast feedback aids the learning process. Ideally employees should have some responsibility for setting their own standards of quantity and quality.

Task identity. Sets of tasks should be separated from other sets of tasks by some clear boundary. Whenever possible a group or individual employee should have responsibility for a set of tasks which is clearly defined, visible and meaningful. In this way work is seen as important by the group or individual undertaking it and others understand and respect its significance.

Task autonomy. Employees should be able to exercise some control over their work. Areas of discretion and decision taking should be available to them.

Other points which warrant attention when designing new systems of work are:

---

* These job-design principles have emerged from the work of a number of leading experts in this field; in particular Emery at the Tavistock Institute in Britain, Thorsrud in Norway and Davis in the USA.

1. The employee must know his job and how well he is performing.
2. He must be able to learn on the job and to go on learning.
3. He needs some social support and recognition in the work situation.
4. He must feel that the job leads to future opportunities.

Only if it is possible to create a task structure which provides employees with an acceptable level of job satisfaction should the chosen solution be accepted.

Step 7: accept the best possible socio-technical system

When the evaluation of the chosen socio-technical solution is complete, it can finally be selected as the best possible socio-technical solution.

However, it is likely that even this solution does not meet all the technical and human needs which were specified at the beginning, although it is the best one available. Any needs which have emerged from your analysis but which have not been met by this solution should now be considered, to see if they can be satisfied in some other way. For example, alterations in personnel policies may enable social needs such as a desire for better promotion opportunities to be achieved.

By following through the steps set out above, the diagnosis and socio-technical systems-design stages of the ETHICS method can be carried out. The further stages needed for the full implementation of a new work system were described in Chapter 2.

The next chapters provide case-study examples of the steps outlined in this chapter, so that you can try out the ETHICS method for yourself. The first exercise involves redesigning a shop-floor work system without any major change in technology. The second and third exercises are the redesign of clerical work systems with, in each case, a new computer system acting as a technical catalyst.

4 The ETHICS method — Exercise 1

BABYCARE LTD — FOOD PACKAGING DEPARTMENT

Background information

In this exercise you are asked to do the following:

1. Use the ETHICS method to make an analysis of the human problems which exist in the Food Packing Department.
2. Use the ETHICS method to produce a socio-technical solution which removes existing problems.

   Set out your solution diagramatically showing:

   - Any changes in the nature or arrangement of machinery.
   - Any changes in workflow.
   - Any changes in group social structure.
   - Any changes in group or individual task structures.
3. Describe changes you propose making to conditions of work, control systems, supervisory style and any other relevant factors.
4. Set out the implications of your proposed work system for recruitment, training and employee development.
5. Prepare a strategy to convince a factory manager who is only interested in meeting production targets that change will be worth while.
6. State briefly what alterations you would make to your proposed socio-technical solution if you were designing an entirely new Food Packing Department.

The problems of conveyor-belt work

From the employee's point of view conveyor-belt work has two important characteristics; it is mechanically paced and it is highly repetitive. Research undertaken in Europe and in the United States suggests that in most factory situations these

Table 4.1 Food Packing Department, Babycare Ltd

Product	Baby food. Milk powder.
Operation	Packing this into bags, which are placed in cardboard cartons, which are wrapped in cellophane.
Production system	Mechanically paced conveyor belts. There are five lines.
Manning	Teams of 16 women per line on the day shifts (14 packers, 2 weighers). Teams are smaller on the evening shift and only two or three lines are in use.
Number of workers	162, (142 grade-3 packers, 18 grade-2 weighers, 2 grade-1 chargehands). These are all women.
Shifts	Are day and evening. 120 women work on the day shift (80 of these are part-time). 40 work on the evening shift (6-9 p.m.), plus the 2 chargehands.
Wage-payment system	A weekly wage according to grade, plus a company production bonus paid annually.
Work method	Girls on each line rotate their jobs once an hour.
Trades unions	The firm is non-unionised except for maintenance workers.

are the two most disliked aspects of work. Mechanical pacing is disliked because it leads to feelings of 'pressure' or of being pushed, and repetition because it leads to feelings of impersonality, a belief on the part of the worker that he has no importance as an individual.

Job satisfaction depends on the development by the worker of favourable attitudes towards his work. Research shows that for repetitive work to be associated with favourable attitudes, two conditions must be present, neither one of which is sufficient by itself. Firstly, the product being worked on and the work process must contain some important perceived attractions. Secondly, in order to maximise these attractions, distractions or parts of the work situation or work process that cause frustration or annoyance to the employee must be eliminated or minimised.

Repetitive work appears to be increasingly unacceptable to the modern worker and reformers argue that, wherever possible, production should be organised in such a manner that mechanically paced conveyor belts need not be used. But despite a statement from the EEC that the paced assembly should not continue, it is still widely used in industry.

The labour turnover problem of the Food Packing Department

This year labour turnover in the Food Packing Department is 73 per cent, last year it was 64.3 per cent. 72 per cent of the women who left the department this year did so within six months of starting work, 46 per cent worked for less than two months. Yet labour that is lost is becoming increasingly difficult to replace, for the firm is located in the London area.

In an attempt to ease the labour shortage the firm is using a considerable number of part-timers. Of the 162 women in the Department, 40 are full-time. The remainder work four-hour shifts in either the morning, afternoon or evening. These shifts are permanent and there is no movement from one shift to another.

30 per cent of leavers are evening-shift workers and this group differs in age and domestic responsibilities from the other shift workers or the full-timers. They are younger and nearly all have small children. Many have been secretaries or clerks in their previous jobs but most had not worked for some time before joining Babycare.

Labour turnover in Food Packing is considerably higher than in other departments which employ women. The adjoining department is Pharmaceutical Packing and here labour turnover is only 33.9 per cent.

45 of the Department's most recent leavers have been interviewed in their homes and asked why they had left Babycare. Their answers were as follows:

Reasons for leaving	%
Work too fast and demanding.	47
Domestic reasons.	36
Work and working conditions not liked.	11

All the Department's present employees have also completed a questionnaire and a sample has been interviewed to find out their attitudes to the work and work situation. Some of their views and comments and those of the leavers are given in the following sections.

Group stability

Although the packing-department women are unlikely to be moved to other departments, there is considerable movement from one belt to another within the Department. Only a quarter work regularly on the same belt. Nearly a half move frequently and the rest move occasionally. Most of those that move frequently dislike this.

> 'For years I've moved around, it makes you very unsettled. I'm much happier when I'm on my own belt with my own team.'
>
> 'Moving is very aggravating. You come in and don't know where you're going. You get used to the girls, then have to move and are with a lot of strangers.'

One women who had left because of this said,

> 'It was never knowing where you would be, and having to wait each day when you started to be told where to go. You got a feeling of working in isolation and you just didn't care about the job. It's a horrible feeling, and nobody likes it. This is the heart of the problem, you're just a number and you don't feel wanted.'

The women claim that one source of discontent is the practice of introducing new women into an established team; then, when they can do the job, leaving them there and moving two of the original team on to another belt. The Forewoman has introduced this system because she believes that girls should not stay too long on one belt, but should have experience of other lines. It has the result, however, of making new girls a threat to existing groups.

A few girls liked to move, however. As one put it, 'I like being an all rounder. I'm funny, I like to have a go at all the jobs.'

Wages, conditions of work, and supervision

A majority of present and previous employees are satisfied with pay, although some say that Food Packing should get more than other departments as the work is so much harder. Working conditions are accepted as reasonable although there are some comments on temperature and congestion. The Forewoman is well liked and the women appreciate her willingness to listen to personnel problems, and her readiness to help and advise.

Training

No off-the-line training is provided in the Department. Each new girl is placed directly on to the line with a more experienced girl or a charge-hand beside her. As so many girls leave within two months of starting work, a number of questions were asked on how the job had been learnt and how long the learning process had taken. It seems likely that girls are leaving before they have completely mastered the work. One-third of both present and former employees said they had picked up the job by watching other girls, one-third had been shown by another girl. The remainder had been shown, at least initially, by a charge-hand. The majority said that although they had picked up job movements reasonably quickly, it had taken them several months to master the speed.

First impressions of a new worker

This individual had worked on assembly operations previously and so the work was not new to her. Even so, she found her first impressions of the job were unpleasant and disturbing. She was taken to a packing line, given a chair, and left to the mercies of the girl next to her on the line. There was little space between girls on the belt and to a newcomer the work arrived at a very fast rate and in batches, rather than evenly spaced out. This prevented the acquisition of a regular working rhythm. It seemed that girls at the beginning of the line were prone to work fast for a period in order to get together a stock of packets which they would then release in quick succession while they secured a brief rest.

Another factor making the work harder than necessary was the tendency of the machine wrapping the packets in cellophane to break down. When this happened the supervision did not always allow the belt to be stopped, a bottle-neck was created and packets coming up the belt had to be stacked on the floor or on trolleys and re-introduced on to the belt when the cellophane machine was back in operation. Dealing with this backlog of work plus the bags coming up from the beginning of the line threw a very heavy additional strain on the operators. The newcomer found her period on the belt stressful and fatiguing and felt her reactions were probably similar to those of any other new starter.

A positive feature of the work

The work situation does, however, have a more positive side. The firm's chief product, baby food, is one in which the women are very interested and with which they are able to identify. This attitude is typified by the reply of one employee to the question 'What do you like most about your job?': 'It's the feeling that in producing baby food, which is sent all over the world, you are doing something useful and helpful for other people.' This identification with the product is demonstrated by the women's attitude to hygiene. There are no criticisms of the Department's very strict rules on cleanliness and dress. On the contrary, the women seek even stricter standards. There are some complaints that the floor is not always clean and that spotlessly clean cloths are not available for cleaning machinery. Many complaints about the speed of work are accompanied by remarks that this prevents the product being packed as perfectly as the women would like. This inability to achieve high work standards is another frustration arising from machine pacing.

Management attitudes

Management, in this instance the departmental and factory production managers, are emphatic that it is impossible to alter existing production arrangements. Nor can any reduction in the speed at which the belts are run be tolerated. Their attitude is that the line has been work-studied and a correct speed arrived at and that, in any case, the existing speed is essential to meet production targets. They are not convinced of the need for training. The argument is put forward that girls must learn on the belt in order to achieve the required speed, and that there will never be a large enough number of new girls starting at any one time to permit a training belt to be manned. Their opinion is that women leave because they are poor quality labour interested only in short-term financial gains. There is little understanding of, or sympathy with, psychological needs and considerable resistance to change.

Production System

Figure 4.1 shows the production system. The first task is performed by the grade-2 weigher who is also called a 'catcher'. She catches the milk powder in a bag as it comes

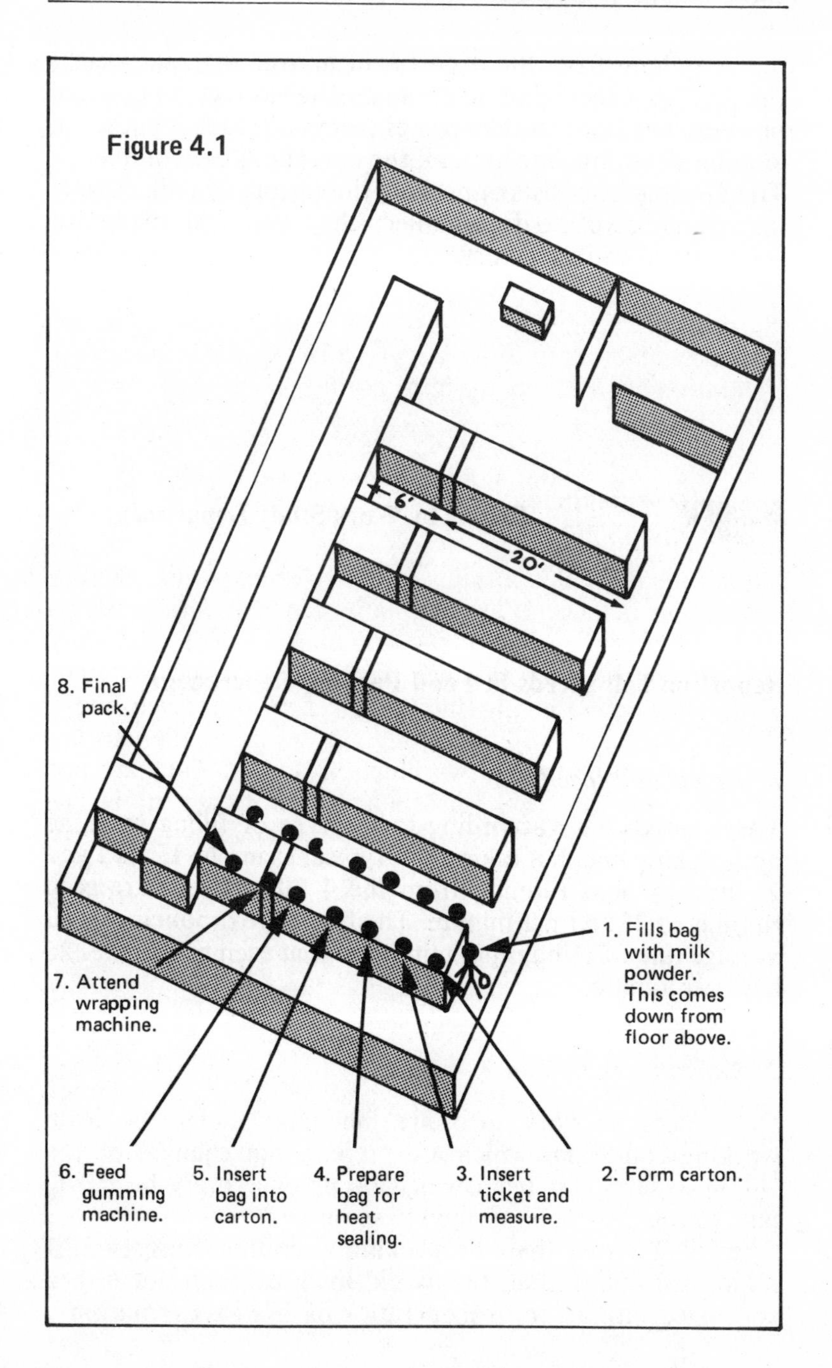
Figure 4.1
6'
20'
8. Final pack.
7. Attend wrapping machine.
1. Fills bag with milk powder. This comes down from floor above.
6. Feed gumming machine.
5. Insert bag into carton.
4. Prepare bag for heat sealing.
3. Insert ticket and measure.
2. Form carton.

down a chute from the department above. She next weighs the bag to check that it contains the correct amount of powder. The product then passes from one grade-3 packer to another down the line until all the operations are completed. The final pack consists of packing the cartons of milk powder in a large cardboard container; they are then ready for shipment from the factory.

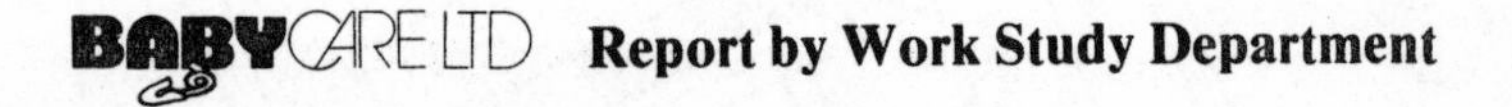

Report by Work Study Department

Report on belt speeds in Food Packing Department

1. Actual belt speeds

These speeds vary according to the type of filling machine and packing line, but taking as a typical example, Lines 1 and 2, the two belts running from the 4 filling heads are each running at 25 feet per minute. The fillers are producing at an average rate of 13 bags per minute so that each belt carries 26 bags per minute.

2. Operator work load

Differences in work load are inevitable under the team working conditions which are present, but changes of job within a particular team which occur once every hour will tend to equalise the work load to some extent.

It will be seen that the packing operation consists of 7 major jobs and that the work load of each job when compared with the control operation of *filling* is as follows:

Form carton	84% of filling operation
Insert ticket and measure	84% of filling operation
Prepare bag for heat sealing	94% of filling operation
Insert bag into carton	97% of filling operation
Feed gumming machine	79% of filling operation
Attend wrapping machine	88% of filling operation
Final pack	76% of filling operation

Report by Personnel Department 1

Some notes on the findings of a research team headed by Dr. K. Murrell on paced work.

Training

This Research Team has established two training facts of relevance to the Babycare situation:

1. Intelligent women are unsuitable for mechanically paced work. They can work extremely fast for a short period, but their speed of work becomes increasingly irregular as they react to the boredom of the task.
2. If women are put straight away onto a paced job, this freezes their rate of work and they are later unable to increase their speed. Women who are taught to perform tasks in an unpaced situation and then are transferred to a mechanically paced belt, can achieve much higher speeds.

The spacing of packers

The congestion of many of the Babycare packing lines and the proximity of each packer to the next, was referred to earlier. Experiments at Bristol University into the most accurate method of relating belt speeds to employee capacity have produced the following conclusion.

If the speed of a belt is arrived at by noting only the time taken by the worker to complete movements, the belt will always be travelling too fast. Three factors need to be taken into account for a correct rate: (i) the speed of movement, (ii) the worker's variability in speed over a period of time, (iii) the 'tolerance' of the task. (Tolerance = the proportion of the task cycle time the packet is available on the belt to be picked up by the worker.) The Bristol research indicates that 'tolerance' is a critical variable and should be taken into account both when deciding the rate at which work should pass along a belt and the spacing of the operatives. The tighter the tolerance of a job, the harder it becomes.

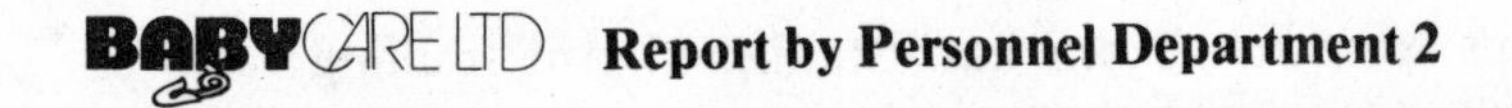

Report by Personnel Department 2

The influence of assembly-line organisation on group output and morale

Some notes on an investigation carried out at the Eindhoven Plant of Philips

These experiments established that the length of the assembly line and the spacing of the operator were crucial factors in the quantity and quality of work produced. The situation investigated was a line used for the assembly of television sets and so there is not an exact parallel with the Babycare situation.

Points from the research report

Production losses due to waiting time

A factor affecting output was the amount of waiting time. This occurred through the differences in average speed between the workers (there are fast and slow workers), and the fact that a worker can never carry out his operations at a constant speed. One moment he completes his operations in a short time, the next moment it takes him longer. Even if workers all work at the same *average* speed, waiting times still occur as a result of individual speed variations per worker. A line can never travel faster than the worker with the longest average operation time. Accordingly, everyone has to wait for this worker.

The investigation showed that the differences in average operation time were not caused by differences in standard times determined by the work study specialist for the various places, but rather by differences in speed between the workers. The differences in standard times were much smaller than speed differences. (The standard times are between plus or minus 8 per cent the average speed of the various workers between plus or minus 25 per cent).

The problem of 'balancing' a line *is not the division of tasks on the basis of official standard times, but rather the adaptation of these tasks to the difference in speed between the workers.*

This problem tends to be greater the longer the line, for it is always the slowest worker who determines the speed of the line. A larger group is more likely to have a worker with a strongly diverging speed than a smaller group.

Even in the case of an ideally balanced line, where each worker on an average takes the same time at his task, waiting times occur because the worker never carries out his operation at a constant speed. Time observations showed that the variation in individual operation times was much greater than expected. If the average individual operation time of a worker is 100 seconds, 99 per cent of his time will be between 75 and 125 seconds.

In order to reduce waiting time the research team experimented with the spacing of workers, with the introduction of 'buffer' stocks into lines and with long and short lines.

By means of job simulation on a computer it was possible to test the effect on output and quality of setting up assembly lines without intervening spaces, and assembly lines with enough space for one television set between each worker. The effects of spacing on lines of 5, 10, 20, 50 and 100 workers were estimated.

It can be seen from the graph in Figure 4.2 that the loss-percentage in a line without spaces is about four times bigger than that in a line provided with a space for one set between each worker. Losses increase as the length of the line is increased.

Summary

Waiting time due to variations in the speed of work between one worker and another, and variations in an individual worker's own speed can be reduced by providing spaces in which a product can be 'parked' for a short time. Short lines with spaces have the least waiting time.

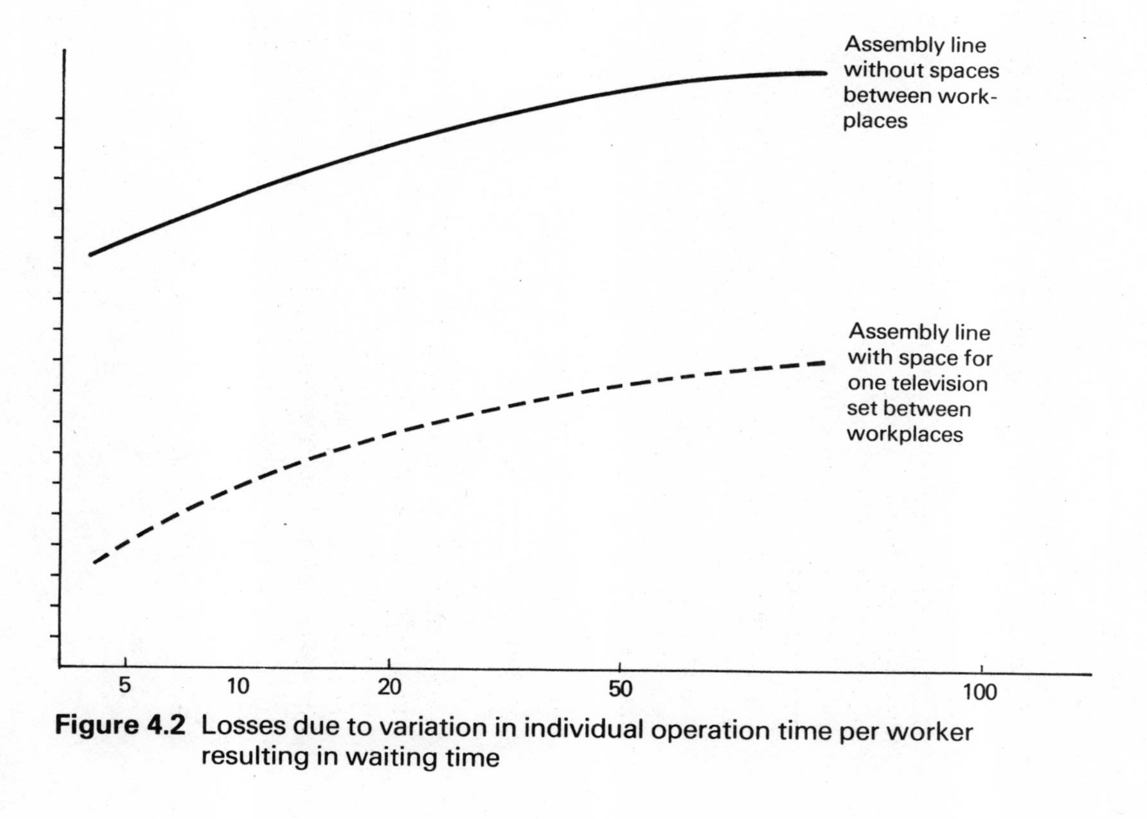

Figure 4.2 Losses due to variation in individual operation time per worker resulting in waiting time

BABYCARE LTD

The Exercise

MAY WE FIRST REMIND YOU OF THE STEPS YOU WILL BE COVERING IN YOUR ANALYSIS AND DESIGN PROCESSES? LOOK ONCE AGAIN AT THE DIAGRAM ON THE NEXT PAGE WHICH YOU HAVE ALREADY SEEN EARLIER IN THIS BOOK.

THE STRUCTURE OF THE EXERCISE WILL LEAD YOU FROM ONE STEP TO THE NEXT. AFTER EACH STEP WE HAVE SET OUT OUR APPROACH TO THAT STAGE OF THE PROBLEM. COMPARE THIS WITH YOURS BUT DO NOT LET IT INFLUENCE YOU TOO MUCH. THERE ARE NO RIGHT OR WRONG ANSWERS TO THE EXERCISE. THE ANSWER YOU ARRIVE AT WILL DEPEND ON YOUR ANALYSIS OF THE PROBLEMS AND THE OBJECTIVES YOU SET YOURSELF.

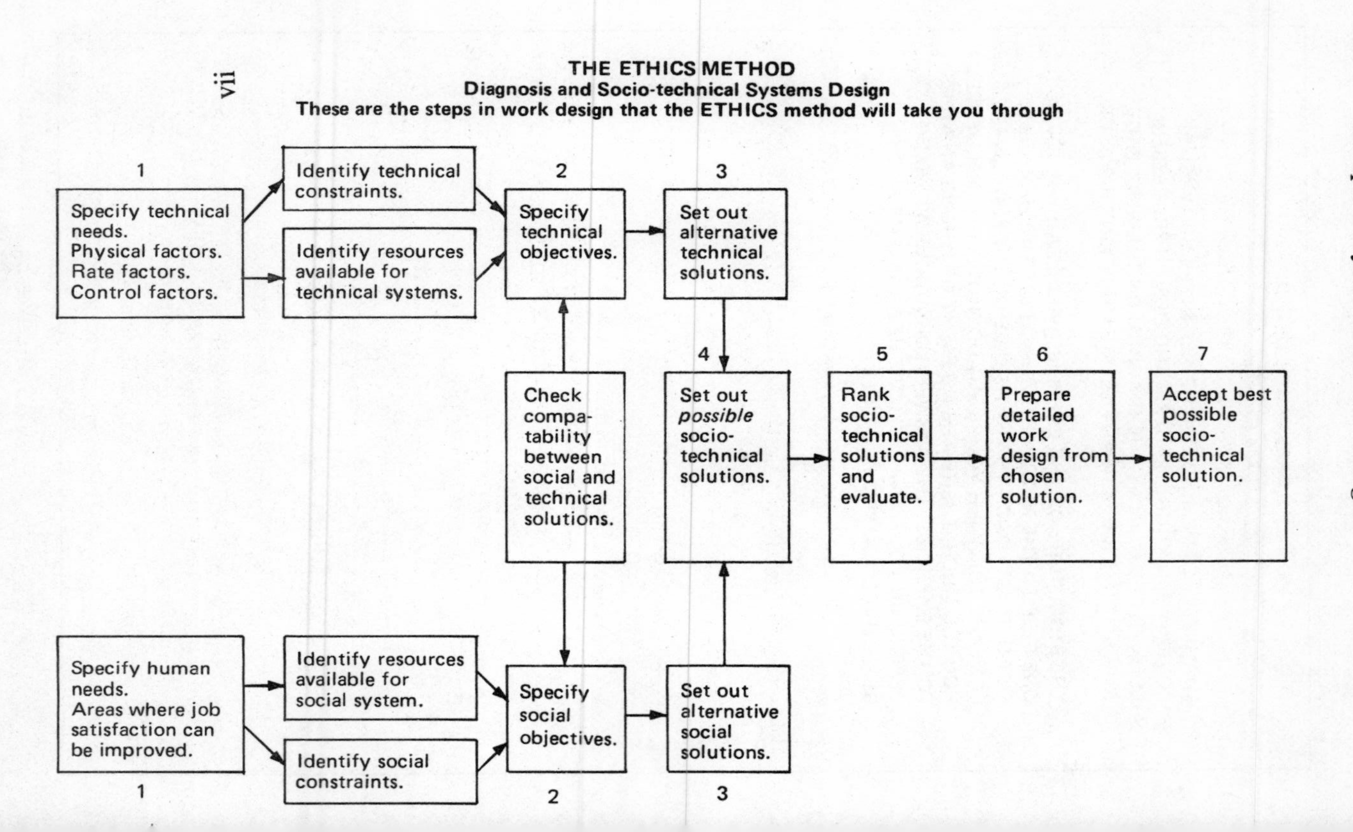
vii
THE ETHICS METHOD
Diagnosis and Socio-technical Systems Design
These are the steps in work design that the ETHICS method will take you through
1
Specify technical needs. Physical factors. Rate factors. Control factors.
Identify technical constraints.
Identify resources available for technical systems.
2
Specify technical objectives.
3
Set out alternative technical solutions.
Check compa-tability between social and technical solutions.
4
Set out possible socio-technical solutions.
5
Rank socio-technical solutions and evaluate.
6
Prepare detailed work design from chosen solution.
7
Accept best possible socio-technical solution.
Specify human needs. Areas where job satisfaction can be improved.
1
Identify resources available for social system.
Identify social constraints.
Specify social objectives.
2
Set out alternative social solutions.
3

BABYCARE LTD

STEP 1 REQUIRES YOU TO MAKE A DETAILED DIAGNOSIS OF JOB-SATISFACTION NEEDS.

BABYCARE LTD

STEP 1 DIAGNOSIS

Specify technical requirements and human needs.

Technical requirements

Physical factors	To produce a work system which is efficient and economical to run. This implies a steady production output, together with an absence of factors likely to cause work stoppages or breakdowns.
Rate factors	Orders are at present greater than production capacity. Rate and volume of work are therefore key factors.
Control factors	Hygiene must be rigorously controlled in view of the nature of the product.

Human needs

Human needs in Babycare can be identified by examining the questionnaire data on the next six pages.

BABYCARE LTD

Workers' Questionnaire

This is a simplified version of the questionnaire that was used in Babycare. The data are presented under our five job-satisfaction headings.

1. If more than two-thirds of respondents agree with a positive statement this is a desirable feature and should, if possible, be incorporated into any redesign of work. Similarly, if more than two-thirds disagree with a negative statement (e.g. I do not like my present job) this is a desirable feature of the work.

2. If more than two-thirds of respondents disagree with a positive statement (e.g. I find my work extremely interesting) this is an undesirable feature and steps should be taken to correct the problem area. Similarly, if more than two-thirds agree with a negative statement this is an undesirable feature of work.

3. Answers falling between these limits can either be ignored or defined as indicating problems, as you wish.

The questionnaire shows the fit between the work needs of women production workers on the Babycare packing line and the extent to which these needs are being met with the present work system and personnel policy. It also shows to some extent how workers, ideally, would like their work needs to be met.

*indicates the workers' response.

BABYCARE LTD

Workers' Questionnaire

The knowledge 'fit'

	More than two-thirds agree	*More than two-thirds disagree*	*Answers falling between these limits*
1. My skills and knowledge are fully used in my present job.		*	
2. I find my work extremely interesting.		*	
3. I do not like my present job.			*
4. I should like a different job in Babycare.			*
5. I should like to be doing a more difficult job.	*		
6. Babycare does not provide enough opportunities for me to learn new things and develop my talents.	*		
7. I should like the opportunity of learning and doing a more challenging kind of job.	*		

BABYCARE LTD

Workers' Questionnaire

The psychological 'fit'

Statement	*More than two-thirds agree*	*More than two-thirds disagree*	*Answers falling between these limits*
1. Status is important to me, I like to be respected.			*
2. Working for Babycare gives me a feeling of status.		*	
3. I carry considerable responsibility in my job.		*	
4. I should like to carry even more responsibility.			*
5. If I do good work, I feel management recognises this.		*	
6. I should like my work to receive more recognition.	*		
7. My job is a very secure one.		*	
8. I enjoy the opportunity for making friends which my job provides.	*		
9. There are not sufficient opportunities for promotion in this department.	*		
10. I should like to get promotion soon.			*
11. I want to achieve a great deal and get to the top		*	
12. I get my feelings of achievement from doing a good job.	*		

BABYCARE LTD

Workers' Questionnaire

The efficiency 'fit'

	More than two-thirds agree	*More than two-thirds disagree*	*Answers falling between these limits*
1. I could earn more money if I did not work for Babycare.		*	
2. I am paid adequately for the work I do.	*		
3. I have to work too hard in my job.	*		
4. We are expected to be too accurate in our work.		*	
5. Mathods of checking are too lax.	*		
6. Supervision here is strict.			*
7. I like to be left to get on with my work without interference from supervision.	*		
8. My supervisors give me all the help I need.	*		
9. The bonus system is fair to everyone. (This is a company production bonus paid annually.)	*		
10. I get all the information I need to do my job.	*		

BABYCARE LTD

Workers' Questionnaire

The task-structure 'fit'

Statement	*More than two-thirds agree*	*More than two-thirds disagree*	*Answers falling between these limits*
1. There is not much variety in my work.	*		
2. I should like a job which was less routine.	*		
3. There is little scope for me to use my own initiative.	*		
4. I should like a job which expected me to use my initiative more.	*		
5. In my job I have to rely on my own judgement and take decisions.		*	
6. It would be better for me if I could take more decisions without having to ask someone first.			*
7. There is a lot of pressure in my job.	*		
8. I should like less pressure in work.	*		
9. I have clear-cut work targets and know when I achieve these.	*		
10. I should like clearer targets to aim at.			*
11. How well I can do my work depends on my co-workers.	*		
12. It would be better for me if I could do my work without depending on others.	*		
13. I can organise and carry out my work the way I want.		*	
14. I should prefer more freedom to plan my work myself.	*		
15. I believe my job to be important and others recognise its importance.	*		
16. I should prefer a job which made a larger and more important contribution to the work of the department.	*		

BABYCARE LTD

Workers' Questionnaire

The ethical 'fit'	*More than two-thirds agree*	*More than two-thirds disagree*	*Answers falling between these limits*
1. Top management in Babycare is too ruthless.		*	
2. Managers and workers in Babycare are very friendly.	*		
3. Babycare puts production above the interests of its employees.	*		
4. Babycare looks after the welfare of its staff very well.	*		
5. A person's character and experience count for more here than formal qualifications.			*
6. I believe that character and experience are more important than qualifications.	*		
7. Communication in this department is very poor.		*	
8. In my section, we are told as much as we want to know about day-to-day matters.	*		
9. Babycare senior management is out of touch with the way the workers feel.	*		
10. There should be better communication between senior management and our department.	*		
11. We have sufficient say in the way Babycare is run.			*
12. Workers should be consulted more about major decisions that are to be made.	*		
13. We cannot sufficiently influence decisions about changes which are made in the way we do our work.	*		

BABYCARE LTD

Workers' Questionnaire

Face-to-face interviews were held with a sample of the women and set out below are verbatim comments which illustrate problems related to the task structure 'fit'.

Work stability

They were asked if they regularly worked with the same group of girls. This emerged as a problem area for there was considerable movement from one belt to another within the department:

Movement from belt to belt	%
Has to move occasionally.	37
Moves frequently.	31
Moves very little.	27

Illustrative comment

'Movement is very aggravating. You come in and don't know where you are going. You get used to the girls, then have to move and are with a lot of strangers'.

The women said that one source of discontent was the practice of introducing new women into an established team, then when they could do the job, leaving them there and moving two of the original team onto another belt.

BABYCARE LTD

Workers' Questionnaire

Face-to-face interviews (continued)

Difficulty of work

The women were asked if they found the work easy or difficult.

Answers	%
Difficult in parts and at times.	27
Difficult at first, but easy now.	23
Easy except for the speed.	27
Easy.	23

Difficulty of work

Illustrative comment

'It's very, very fast when you first come. I've never worked so hard for my money.'

'Some days the job goes sailing along with no difficulties. On other days I could down tools, go in the office and ask for my cards.'

Complaints about the excessive speed of the work were made in answer to subsequent questions. Many said that the work was difficult at the existing rate of 15 packets per minute but that it would not be difficult at 12 or 13 packets per minute.

Women complained that not only was the rate of work very fast, speed was not constant. If a hold-up higher up the belt caused a backlog of packets to accumulate, then work was extra fast until these had been cleared. On the evening shift the belts are never stopped even if the wrapping machine breaks down.

The short length of the belts means that the women have to work closely together and cannot be spaced out along the belt. This increases work difficulty, particularly if a trainee is added to the numbers on the belt.

BABYCARE LTD

Workers' Questionnaire

Face-to-face interviews (continued)

Training

Length of time it took to learn the job

	%
Less than 1 month.	41
1–4 months.	44
Still learning.	8

Many of those who said less than one month added that this did not mean that they had reached the required speed by that time.

37% of the grade-3 packers said that they had been shown how to do the job by another girl on the belt. 34% said that they had not been shown by anyone but had been 'left to get on with it'. The remainder said that they had been taught by the weigher or charge-hand.

Wages

There was no incentive scheme in the department although the firm had a profit-sharing scheme. Women's wages were standard throughout the firm and related to grade.

60% said their wages were satisfactory. 17% thought that Food Packing should receive more than other departments.

'The speed of work is so great and clothes have to be cleaned often because of the powder.'

Conditions of work

Attitudes to these varied according to whether the women were full-time or shift workers. 71% of the evening shift workers thought their working conditions were 'quite good'.

60% of the afternoon shift workers thought conditions were 'quite good', but 30% thought the department was too hot in summer and too cold in winter.

50% of the morning shift thought conditions were quite good. 25% thought the department was overcrowded.

Only 35% of the full-time workers thought conditions 'quite good'. 50% complained of overcrowding and congestion. 15% complained of the temperature.

BABYCARE LTD

Workers' Questionnaire

Face-to-face interviews (continued)

First impressions of Babycare

73% of those who had started work in the Food Packing Department said that they had found it a frightening or unpleasant experience.

Illustrative comment

'Terrible, the work was coming down so fast. You've no speed and no-one to help you.'

'I was a nervous wreck, shaking like a leaf. It was nerve-racking.'

'I thought I'd never stick it. I thought I'd never keep the work speed up. It all terrified me, not being used to factory work.'

Attitudes to new workers

These were viewed as a nuisance. They inconvenienced more experienced workers by slowing them down, particularly when they had to be given assistance.

Attitudes to technical maintenance staff

These were seen as both inefficient and a threat. They were slow to make repairs when there was a breakdown and prone to alter the speed of the belt when they restarted the line after a breakdown. They did not work for Food Packing but for a Central Maintenance Department situated some distance away.

Attitudes to supervision

The day-shift forewoman was well liked but the evening-shift charge-hands were seen as poor at their job.

BABYCARE LTD

ON THE BASIS OF A CAREFUL ANALYSIS OF THE QUESTIONNAIRE DATA AND ILLUSTRATIVE COMMENTS, INDICATE ON THE NEXT THREE PAGES THOSE ASPECTS OF THE EXISTING WORK SYSTEM THAT YOU REGARD AS SATISFACTORY AND WOULD WISH TO RETAIN IN ANY REDESIGN OF THE WORK OF THE DEPARTMENT. INDICATE ALSO THOSE YOU REGARD AS UNSATISFACTORY AND WOULD WISH TO IMPROVE. IF YOU GET ANY IDEAS ON POSSIBLE WAYS OF MAKING IMPROVEMENTS AS YOU WORK DOWN THE LIST, WRITE THESE IN THE RIGHT-HAND COLUMN SO THAT YOU DO NOT FORGET THEM.

(Asterisk or make written comments.
as appropriate)

BABYCARE LTD
Analysis of Human Needs

	Satisfactory aspects	*Should be incorporated in new system?*	*Unsatisfactory aspects*	*Must be improved in new system*	*Suggestions on how this can be achieved*
Knowledge 'fit'					
Use of knowledge					
Self-development					
Psychological 'fit'					
Status					
Responsibility					
Recognition					
Job security					
Social relationships					
Promotion					
Achievement					

BABYCARE LTD
Analysis of Human Needs

	Satisfactory aspects	*Should be incorporated in new system?*	*Unsatisfactory aspects*	*Must be improved in new system*	*Suggestions on how this can be achieved*
Efficiency 'fit'					
Salary					
Amount of work					
Standards					
Controls					
Supervision					
Task structure 'fit'					
Work variety					
Initiative					
Judgement and decisions					
Pressure					
Targets					
Dependency on others					
Autonomy					
Task identity					

BABYCARE LTD

Analysis of Human Needs

	Satisfactory aspects	*Should be incorporated into new system?*	*Unsatisfactory aspects*	*Must be improved in new system*	*Suggestions on how this can be achieved*
Ethical 'fit'					
Company ethos					
Communication					
Consultation					

Please summarise your analysis and enter your summary into the box below, noting the most important areas that require improvement.

Summary of required improvements	

BABYCARE LTD

THERE ARE NO COMPLETELY RIGHT OR WRONG ANSWERS, BUT ON THE BASIS OF THE INFORMATION YOU HAVE BEEN GIVEN, YOUR ANALYSIS OF HUMAN (JOB-SATISFACTION) NEEDS MAY BE SIMILAR TO THE FOLLOWING.

(Asterisk or make written comments as appropriate)

BABYCARE LTD

Analysis of Human Needs

	Satisfactory aspects	*Should be incorporated in new system?*	*Unsatisfactory aspects*	*Must be improved in new system*	*Suggestions on how this can be achieved*
Knowledge 'fit'					
Use of knowledge			Skills associated with coping with the speed of work are difficult to acquire. Few other skills or knowledge are required.	*	Through the re-organisation of work. This can eliminate paced work and substitute larger jobs requiring higher level skills and more decision taking.
Self-development			There are no opportunities for self-development	*	Through the re-organisation of work plus promotion policies which provide a route from the shop-floor to office jobs or supervision.
Psychological 'fit'					
Status			Not a serious problem but could be improved.		Larger jobs should provide more status.
Responsibility			See above.		Larger jobs could provide more responsibility.
Recognition			Women want more of this.	*	More involvement in departmental decisions. Altered management style.
Job security	*		Jobs are secure but women do not think they are.		
Social relationships	These are already good.	Yes.			A greater degree of group stability would assist this.
Promotion	No strong desire for promotion by the majority.				See self-development. A route upwards should be provided for those who want it.
Achievement	*	Yes.			Through continuation of high-quality standards.

BABYCARE LTD
Analysis of Human Needs

	Satisfactory aspects	*Should be incorporated in new system?*	*Unsatisfactory aspects*	*Must be improved in new system*	*Suggestions on how this can be achieved*
Efficiency 'fit'					
Salary	*	Yes.			No deterioration in existing effort-reward bargain.
Amount of work			This is related to the speed of work.	*	Eliminate mechanical pacing.
Standards	*	Yes.			No deterioration in quality standards.
Controls			Technical controls are poor.		Changed methods of work would alter existing technical controls.
Supervision	*	Yes.	Some problems with evening-shift supervision.		Give evening shift more support.
Task structure 'fit'					
Work variety			Work too repetitive and routine.	*	Through a reorganisation of the department and a restructuring of the tasks of individual workers; together with some new policy on advancement, communication and consultation.
Initiative			Little or no initiative is required.	*	
Judgement and decisions			There are few opportunities for this.	*	
Pressure			High pressure due to mechanical pacing and speed or work.	*	
Targets	*	Yes.			
Dependency on others			One job is totally dependent on others.	*	
Autonomy			No autonomy for individual.	*	
Task identity	This comes from product not job.	Yes.	Jobs too small.	*	

BABYCARE LTD

Analysis of Human Needs

	Satisfactory aspects	*Should be incorporated into new system?*	*Unsatisfactory aspects*	*Must be improved in new system*	*Suggestions on how this can be achieved*
Ethical 'fit'					
Company ethos	There is approval of company philosophy and values . . .	Yes.	but, company is seen as too production-minded.		Reorganisation of work would reduce production 'pressure'.
Communication	Good within the department.	Yes.	Poor between department and higher levels in company.		More involvement of workers in policy decisions concerning the department.
Consultation			There is a desire for more consultation.	*	See above.
Summary of required improvements	There is a need for work which is less stressful and more varied, interesting, skilled and challenging. A situation needs to be created in which workers can realise their potential while contributing to company objectives. Workers need better and more effective communication and consultation. Some want opportunities for advancement.				

BABYCARE LTD

STEP 2 CONSISTS OF IDENTIFYING THE TECHNICAL AND HUMAN CONSTRAINTS IN THE BABYCARE SITUATION; THAT IS, FACTORS WHICH MAY MAKE CERTAIN SOLUTIONS NON-VIABLE. IT ALSO CONSISTS OF IDENTIFYING EXISTING RESOURCES WHICH WILL ASSIST YOU IN YOUR EFFORTS TO INTRODUCE MAJOR CHANGE.

BOTH THESE CONSTRAINTS AND RESOURCES ARE DESCRIBED FOR YOU.

YOU ARE ALSO ASKED TO SPECIFY BROAD TECHNICAL OBJECTIVES RELATING TO INCREASING EFFICIENCY AND HUMAN OBJECTIVES RELATED TO INCREASING JOB SATISFACTION. TECHNICAL OBJECTIVES ARE GIVEN YOU. HUMAN OBJECTIVES SHOULD BE DERIVED FROM YOUR ANALYSIS OF HUMAN NEEDS.

BABYCARE LTD

STEP 2 SOCIO-TECHNICAL SYSTEMS DESIGN FOR OVERCOMING EXISTING PROBLEMS

Identify constraints

Identify technical constraints on the design of the system

Improvements must be carried out in the existing department. There is some money available for this improvement, but not sufficient for automation to be a viable solution. Semi-automatic machinery would also not be acceptable on any large scale.

There must be no increased costs or decrease in production in the day-to-day running of the department.

Identify social constraints on the design of the system

The existing labour force must be accepted. This has a nucleus of responsible, conscientious, long-service women who are highly identified with the company. But it has a high labour turnover amongst new starters. (Labour turnover is 73%. 72% of women leaving do so within 6 months of starting work. 46% of these work for less than 2 months.)

The Personnel Manager is actively interested in improving the situation in Food Packing, the Factory Manager is a technical man with little sympathy for, or understanding of, human needs.

This is a non-unionised firm except for some skilled groups such as maintenance.

BABYCARE LTD

Identify resources

Identify resources available for the technical system

Expertise A sympathetic work-study engineer will help to test out proposals for altering the existing technical system.

Technology The Forgrove wrapping machines, which place cellophane around the packets of baby food, are very prone to breakdown. Better models are now on the market.

Finance Top management will allocate a budget for developments which it considers likely to lead to improvement.

Identify resources available for the social system

Top management support The Managing Director will support change if this improves both job satisfaction and efficiency.

Recognition of costs The existing rate of labour turnover is causing serious production problems. Lines often cannot be run because of a shortage of staff. The labour market is tight and becoming tighter, with the result that labour lost cannot be replaced.

Information There is now a great deal of research evidence that this kind of assembly-line operation is expensive to run because of its human disadvantages, and that alternative production arrangements work better and cost less.

BABYCARE LTD

Specify broad efficiency and job-satisfaction objectives

Technical (efficiency) objectives

1. To meet current market requirements.
2. To decrease expenditure per unit of quality produced or, at least not to increase it.
3. To create an efficient production situation which can adapt to expanding market requirements.
4. To ensure that a first-class, pure and hygienic product is produced.

Human (job-satisfaction) objectives (These should be derived from your analysis of human needs.)

CHECK THAT YOUR TECHNICAL AND HUMAN OBJECTIVES ARE COMPATIBLE.

BABYCARE LTD

THE HUMAN OBJECTIVES WHICH WE HAVE SET ARE:

Human (job-satisfaction) objectives

1. To secure a stable, motivated work-force with high efficiency and high job satisfaction.
2. To create conditions in which workers can realise their potential while contributing to company objectives.
3. To ensure that the method of production is such that it causes no mental or physical stress

BABYCARE LTD

STEP 3 REQUIRES YOU TO SET OUT A NUMBER OF ALTERNATIVE TECHNICAL SOLUTIONS AND TO SPECIFY THE EFFICIENCY AND JOB-SATISFACTION ADVANTAGES AND DISADVANTAGES OF EACH OF THESE.

BY TECHNICAL ALTERNATIVE IS MEANT A METHOD OF ORGANISING THE PRODUCTION PROCESS: FOR EXAMPLE, A FLOW-LINE SYSTEM, A BATCH SYSTEM ETC.

BABYCARE LTD

STEP 3 SETTING OUT ALTERNATIVE SOLUTIONS

Technical solutions

It may help to break down the technical system into a number of factors when you are working out your technical solutions. These are:

1. *Technology* The kinds of machines and machine layout associated with the system.
2. *Production procedure* The different operations which have to be completed in order to produce the product and the organisation and sequence of these.
3. *Maintenance activities* Activities related to keeping the system running. These are now the responsibility of technical maintenance staff.
4. *Controls* Systems for ensuring that quality, quantity and hygiene standards are maintained.
5. *Technical weaknesses* Aspects of the technical system which are prone to deviate from some desired standard or norm. You should aim to eliminate or reduce these.

Set out below a number of different ways of altering some or all of the above, so as to improve efficiency and achieve your technical objectives. For example, you may be able to change machines or machine layout, although you must bear in mind that there are serious cost constraints on any major replacement of machinery. You may be able to reorganise the work flow or alter other aspects of the production procedures, or change the maintenance or control procedures. If you can overcome existing technical weaknesses this will help to create the efficient technical system which is one of your objectives. For each of your solutions set out its efficiency and job-satisfaction advantages and disadvantages.

BABYCARE LTD

Description of technical solution	Advantages	Disadvantages
	Efficiency advantages	Efficiency disadvantages
	Job-satisfaction advantages	Job-satisfaction disadvantages
	Efficiency advantages	Efficiency disadvantages
	Job-satisfaction advantages	Job-satisfaction disadvantages

BABYCARE LTD

Description of technical solution	*Advantages*	*Disadvantages*
	Efficiency advantages	*Efficiency disadvantages*
	Job-satisfaction advantages	*Job-satisfaction disadvantages*
	Efficiency advantages	*Efficiency disadvantages*
	Job-satisfaction advantages	*Job-satisfaction disadvantages*

BABYCARE LTD

Description of technical solution	*Advantages*	*Disadvantages*
	Efficiency advantages	*Efficiency disadvantages*
	Job-satisfaction advantages	*Job-satisfaction disadvantages*
	Efficiency advantages	*Efficiency disadvantages*
	Job-satisfaction advantages	*Job-satisfaction disadvantages*

BABYCARE LTD

Now do a preliminary evaluation of the alternative *technical* solutions you have put forward.

For each solution ask yourself:

1. Does it achieve the technical requirements which we specified in Step 1?
2. Is it limited in any way by the constraints identified earlier?
3. Are the available resources which were identified earlier adequate to achieve this solution?
4. Does it meet the objectives which were set out earlier?

You can now draw up a short-list of technical solutions which still seem reasonable after they have been examined in this way.

Try and keep your short-list down to three or four solutions. Write out your short-list on the next page.

BABYCARE LTD

Technical solutions short-list

Give brief description

1.

2.

3.

4.

BABYCARE LTD

HERE AGAIN THERE ARE NO COMPLETELY RIGHT OR WRONG TECHNICAL SOLUTIONS. OUR SUGGESTIONS ARE ON THE FOLLOWING PAGES. YOURS MAY WELL BE BETTER. IT IS THE APPROACH THAT MATTERS.

NB These examples are merely to illustrate the method.

BABYCARE LTD

Description of technical solution	*Advantages*	*Disadvantages*
T1. *Change technology* Technology — replace wrapping machines. Production procedures — as now. Maintenance — as now. Controls — as now.	*Efficiency advantages* No disturbance to existing organisation of department. Level of output maintained or increased. New wrapping machines would reduce stoppages and improve a major technical weakness.	*Efficiency disadvantages* High labour turnover and absenteeism still means all lines cannot be manned, with consequent loss of production. Five new machines would cost a great deal of money. Line-balancing will still be a problem.
	Job-satisfaction advantages Few Existing system rapidly screens out women who cannot tolerate the work or keep up with the pace, therefore, some selection of those best likely to adapt does take place.	*Job-satisfaction disadvantages* Boring, repetitive, paced work continues. High labour turnover continues with increasing difficulty in replacing labour in tight labour market.
T2. *Change maintenance responsibility* Give group some *maintenance skills* so that they can look after and repair the wrapping machine, or its replacement. Technology — as now or replace wrapping machine. Production procedures — as now. Maintenance — now to a degree independent of technicians Controls — as now.	*Efficiency advantages* Risk of some long production hold-ups through machine breakdown is avoided. Employees can get the line working again quickly without having to wait for the maintenance technicians to arrive. Avoids cost of major reorganisation. Helps overcome a technical weakness.	*Efficiency disadvantages* As in solution above. The women will only be able to deal with simple breakdowns. Maintenance men will still have to be called for serious breakdowns.
	Job-satisfaction advantages The workers will acquire some new skills and have a new area of work variety and interest.	*Job satisfaction disadvantages* May be socially disruptive as an elite group of women who have more skills than others could be created. Also, does not alter routine nature of the rest of the work or remove stress of mechanical pacing.

NB These examples are merely to illustrate the method.

BABYCARE LTD

Description of technical solution	*Advantages*	*Disadvantages*
T3. *Change technology and production procedures* Technology — replace wrapping machine. Production procedures — introduce buffer-stock system. Maintenance — as in solutions 1 or 2. Controls — as now.	*Efficiency advantages* The introduction of buffer stock may increase production because this form of technical organisation caters for differences in the speed of work of individual workers.	*Efficiency disadvantages* As in solutions (1) and (2). Production may decrease as the sliding of the packet into the buffer-stock bank adds an additional operation to the tasks of each worker. Also a buffer-stock system may make the department even more cramped to work in than it is at present.
	Job-satisfaction advantages A buffer-stock conveyor-belt system means that each worker completes the operations she is responsible for. The work-piece is then slid off the moving belt and added to a bank of work-pieces at the side of the belt. The next worker uses these as her supply of components. This system enables each worker to choose the speed of work most comfortable for her.	*Job-satisfaction disadvantages* Boring and repetitive work continues although the pressure of machine-paced work is eased for the individual.
T4. *Change technology, maintenance and technical control system* Technology — replace wrapping machine. Maintenance — as in solutions 1 or 2. Controls — *workers are able to control the pace of the line.* To slow it down and speed it up through an adjustment mechanism.	*Efficiency advantages* Giving control of the pace of the belt to the group may increase production because this form of technical organisation caters for differences in the speed of work of the line group as a whole; e.g. at different times of the day, or in different room temperatures.	*Efficiency disadvantages* As in solutions (1) and (2). Production may fluctuate during the day, week or month according to room temperature, day of the week, beginning or end of the month, etc.
	Job-satisfaction advantages Giving the workers control over the pace of the belt means that they can alter this to fit their fatigue level; e.g. run it fast at the beginning of the day or in cold weather, slow it down at the end of the day or in hot weather.	*Job-satisfaction disadvantages* Boring and repetitive work continues. This solution does not allow for individual differences in speed of work, it caters only for the needs of the group as a whole.

BABYCARE LTD

Description of technical solution	*Advantages*	*Disadvantages*
T5. *Change technology, production procedures, maintenance and technical controls.* Technology — abandon paced assembly line Production procedures — *reorganise lines as independent groups of work stations without mechanically paced belts* Maintenance — as in solutions 1 or 2. Technical *controls* which determine the pace of work are abandoned once the paced assembly line is abandoned.	*Efficiency advantages* Less vulnerable technical system. If one work station stops this does not affect the others.	*Efficiency disadvantages* Absence of any machine pacing may reduce output. The success of this kind of work system depends on the design of an efficient, reliable materials handling system. Some duplicate tools and fixtures required.
	Job-satisfaction advantages This form of technical work organisation caters for individual differences in speed of work and facilitates the building-up of jobs to incorporate variety, challenge, autonomy and interest.	*Job-satisfaction disadvantages* Some workers may not have the wish or potential skill to adjust to this form of work organisation.

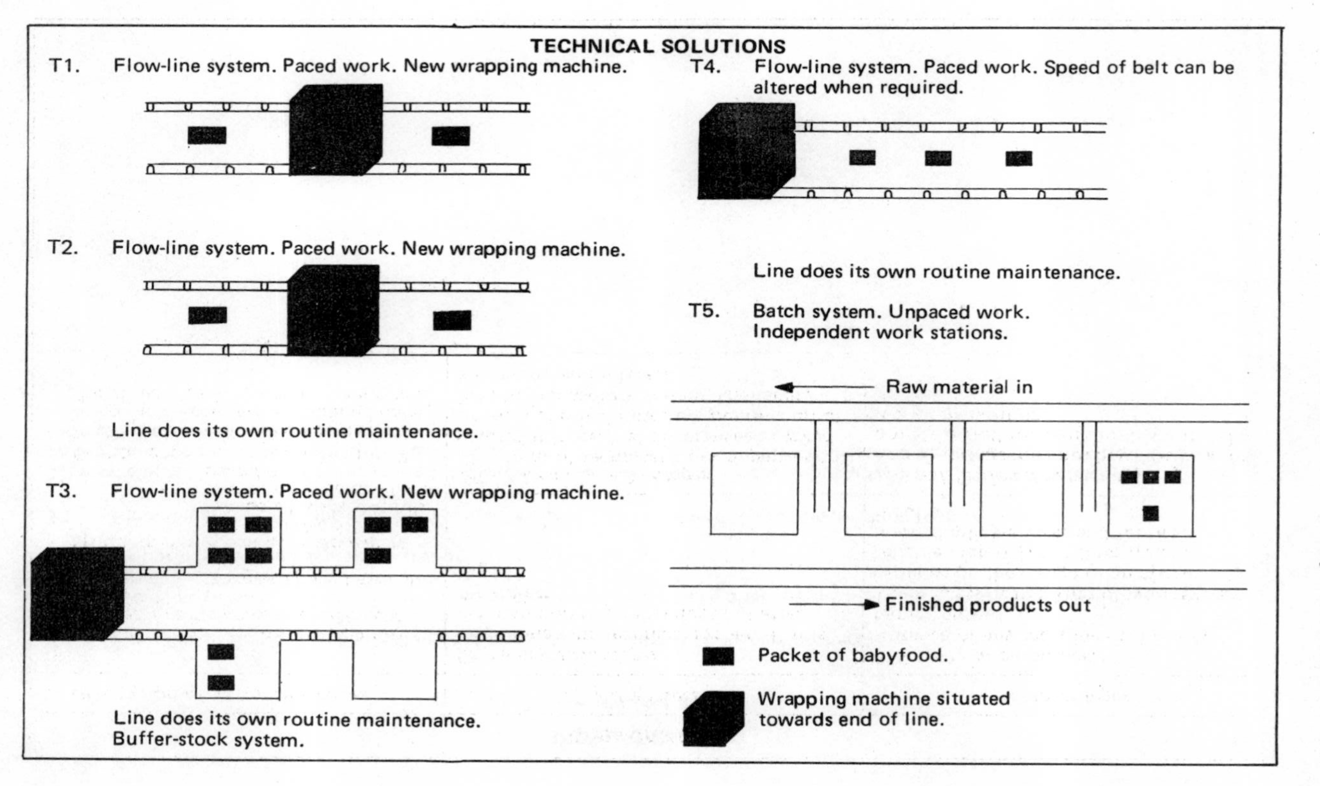
TECHNICAL SOLUTIONS
T1. Flow-line system. Paced work. New wrapping machine.
T2. Flow-line system. Paced work. New wrapping machine.
Line does its own routine maintenance.
T3. Flow-line system. Paced work. New wrapping machine.
Line does its own routine maintenance.
Buffer-stock system.
T4. Flow-line system. Paced work. Speed of belt can be altered when required.
Line does its own routine maintenance.
T5. Batch system. Unpaced work.
Independent work stations.
Raw material in
Finished products out
Packet of babyfood.
Wrapping machine situated towards end of line.

BABYCARE LTD

Technical solution short-list

T3. Flow-line principle. Mechanically paced line with buffer stocks between workers.

T4. Flow-line principle. Mechanically paced line with workers able to alter the speed of the line to meet their fatigue level.

T5. Cell principle. Individual work stations with no mechanical pacing.

T1. Flow-line principle. Line as it is now, mechanically paced but with wrapping machine replaced.

T2 was rejected because it was recognised that giving the women packers responsibility for some routine machine maintenance would cause trades-union problems which would take some time to resolve. This is therefore a solution for the future and may have to be excluded from all the solutions, in the short term.

BABYCARE LTD

STEP 3 SETTING OUT ALTERNATIVE SOCIAL SOLUTIONS

Now proceed to set out alternative social solutions directed at improving job satisfaction. At this stage you may think about these independently of the technical solutions which you have just set out. It may help to break the social system down into a number of factors:

1. The creation of a *work group* structure which is efficient and socially satisfying.
2. The nature and number of tasks which are the responsibility of each work group.
3. The creation of a set of tasks for each *individual* which are interesting, challenging and satisfying.
4. The nature and number of tasks which are the responsibility of an individual worker.
5. *Social controls and feedback.* The possibility of output, quality and hygiene targets being set by the workers themselves.
6. *Feedback* to the workers so that they can monitor their own performance.
7. *Relationships.* The creation of a work situation which facilitates good social relationships.
8. *Decision-taking.* The possibility of providing work groups or individuals with the opportunity to use discretion and make decisions.

At present concentrate on (1), (3), (5), (6). Leave (2), (4), (7) and (8) until Step 7.

BABYCARE LTD		
Description of social solution	*Advantages*	*Disadvantages*
	Job-satisfaction advantages	*Job-satisfaction disadvantages*
	Efficiency advantages	*Efficiency disadvantages*
	Job-satisfaction advantages	*Job-satisfaction disadvantages*
	Efficiency advantages	*Efficiency disadvantages*

BABYCARE LTD

Description of social solution	*Advantages*	*Disadvantages*
	Job-satisfaction advantages	*Job-satisfaction disadvantages*
	Efficiency advantages	*Efficiency disadvantages*
	Job-satisfaction advantages	*Job-satisfaction disadvantages*
	Efficiency advantages	*Efficiency disadvantages*

BABYCARE LTD		
Description of social solution	*Advantages*	*Disadvantages*
	Job-satisfaction advantages	*Job-satisfaction disadvantages*
	Efficiency advantages	*Efficiency disadvantages*
	Job-satisfaction advantages	*Job-satisfaction disadvantages*
	Efficiency advantages	*Efficiency disadvantages*

BABYCARE LTD

When you have set out several alternative ways of organising the Food Packing Department, do a preliminary evaluation of your social solutions as you did for the technical solutions, as follows:

1. Do the solutions meet the social needs which you identified in Step 1?

2. Are any of the solutions constrained by the conditions which you identified earlier?

3. Have you got the resources available to achieve each of the solutions you have set up?

4. Do all the solutions achieve the human objectives which you established earlier?

Having checked all your social solutions against these four criteria, eliminate any about which you are doubtful, and draw up a short-list of the remainder, if possible no more than three or four solutions. Enter your short-list of social solutions on to the right-hand side of the next page. Copy your list of preferred technical solutions which you wrote in on page 94 onto the left-hand side.

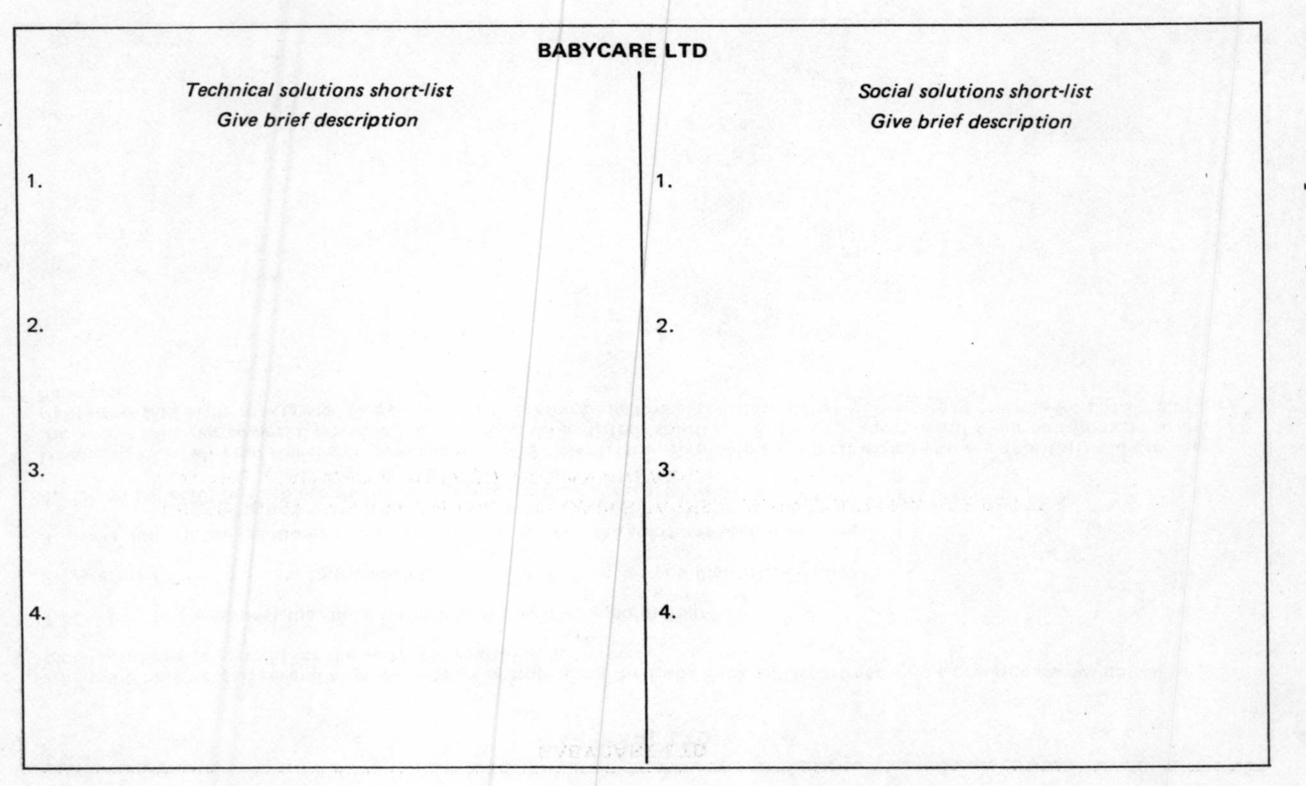

BABYCARE LTD

Technical solutions short-list *Give brief description*	*Social solutions short-list* *Give brief description*
1.	1.
2.	2.
3.	3.
4.	4.

BABYCARE LTD

OUR SUGGESTIONS ARE ON THE FOLLOWING PAGES. AGAIN YOURS MAY BE BETTER.
YOU ARE LEARNING A METHOD, NOT AN ANSWER.

BABYCARE LTD

Description of social solution	*Advantages*	*Disadvantages*
S1. *Work organisation.* Line-type work structure is retained. *Individual worker's tasks.* Each worker is assigned one operation as a full-time job (with work rotation if desired). *Social controls.* Targets set and monitored by supervision. *Feedback.* From supervision if group fails to meet targets.	*Job-satisfaction advantages* Undemanding, secure form of organisation. Everyone knows what they should be doing. Workers have a very limited area of responsibility. Job is easy to learn. *Efficiency advantages* A very easy kind of work system to control. Workers have little discretion.	*Job-satisfaction disadvantages* Work is routine and repetitive. If workers do not like the system, the firm can easily slip into a vanishing labour-force syndrome—with high labour turnover in a tightening labour market. This has already happened at Babycare. *Efficiency disadvantages* Technically vulnerable to breakdown. Line operates at pace of slowest worker because of the task interdependence.
S2. *Work organisation.* Line-type work structure is retained. *Individual worker's tasks.* Each worker is assigned a group of operations as a full-time job. *Social controls.* As above. *Feedback.* As above.	*Job-satisfaction advantages* As above, plus some work variety, although this may take the form of job enlargement rather than job enrichment. *Efficiency advantages* As above.	*Job-satisfaction disadvantages* As above, although less so if group of operations has some with a high skill content. *Efficiency disadvantages* As above. It may be difficult to balance this kind of line without a buffer stock between each worker.

BABYCARE LTD		
Description of social solution	*Advantages*	*Disadvantages*
S3. *Work organisation.* Line or circle. The work group is given all the operations necessary to complete a job (make a product). They can divide the tasks among themselves, informally. *Social controls.* Targets are set and monitored by group and supervision co-operatively. *Feedback.* To group. Information on extent to which targets have been achieved.	*Job-satisfaction advantages* The group becomes responsible for how the work is carried out. This approach provides a good learning environment, for all group members can do all the tasks. *Efficiency advantages* Produces competent, versatile group capable of organising its own activities.	*Job-satisfaction disadvantages* The group may not use its work allocation responsibilities very wisely. Some workers may take all the interesting jobs. A group may not be benevolent to its own members. *Efficiency disadvantages* Not easy to control. Responsibility and discretion may be transferred from supervision to work group.
S4. *Work organisation.* Line, circle or no group structure with independent work stations. *Individual worker's tasks.* Each worker is assigned all the operations required to complete a job. *Social controls.* Targets set by group or by individual worker. *Feedback.* To group or individual. As above.	*Job-satisfaction advantages* Worker has satisfaction of doing a 'whole job' and has considerable work independence. *Efficiency advantages* Poor work can be easily traced to the worker responsible. Highly skilled workers are developed.	*Job-satisfaction disadvantages* Workers may lose the feeling of belonging to a group and work as isolated individuals. *Efficiency disadvantages* There may be difficulty in achieving the required work output.

BABYCARE LTD		
Description of social solution	*Advantages*	*Disadvantages*
S5. *Work organisation.* Square, rectangular or circular group with independent work stations. *Individual worker's tasks.* Each worker is assigned all the operations required to complete a job. *Group tasks.* The group is given responsibility for ancilliary tasks of various kinds. *Targets.* Set by group. *Feedback.* To group to enable it to do its own monitoring.	*Job-satisfaction advantages* Group has high responsibility. Individual has satisfaction of completing 'whole job'. Plenty of work variety. *Efficiency advantages* Group is almost entirely self-controlled. Has little dependence on supervision. If motivated group members will be high producers.	*Job-satisfaction disadvantages* There may be too much variety and responsibility for some group members. *Efficiency disadvantages* As for efficiency advantages. If not motivated group members can be low producers.

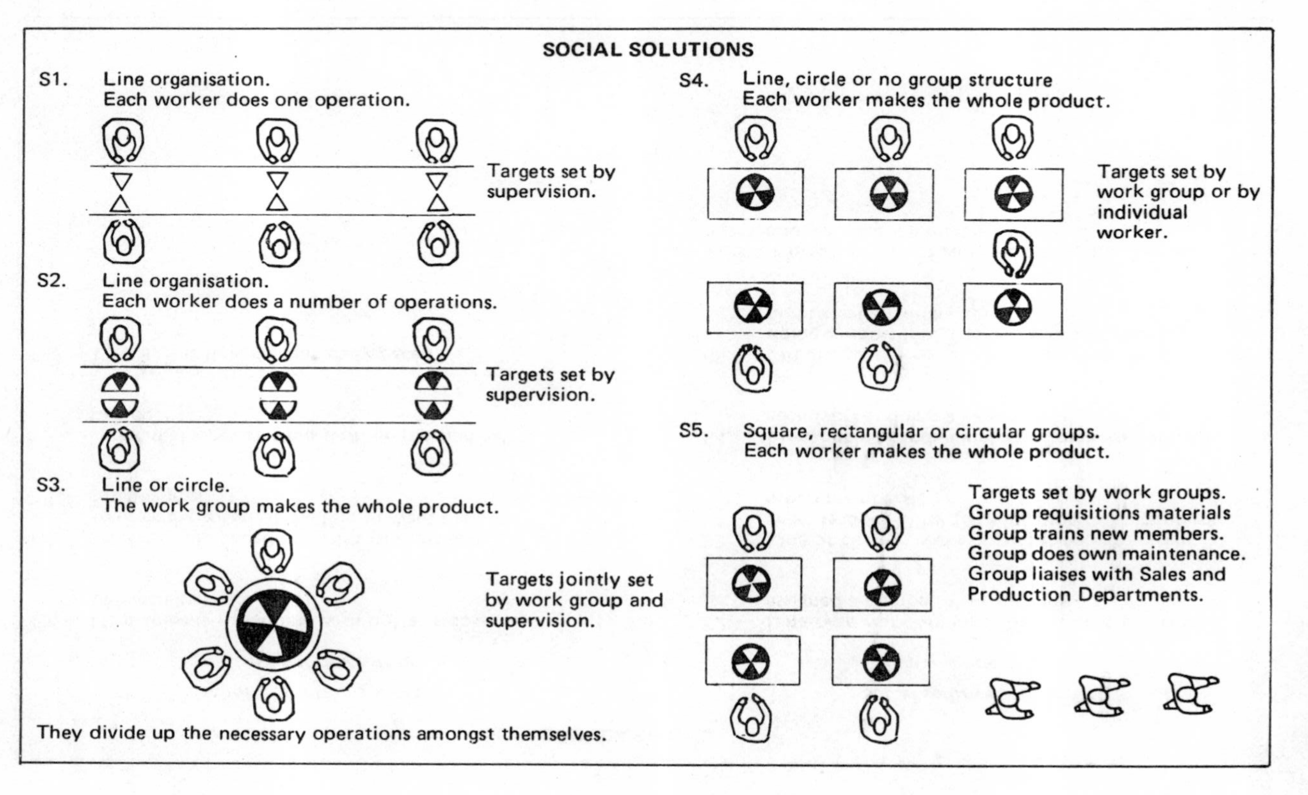
SOCIAL SOLUTIONS
S1.
Line organisation.
Each worker does one operation.
Targets set by supervision.
S2.
Line organisation.
Each worker does a number of operations.
Targets set by supervision.
S3.
Line or circle.
The work group makes the whole product.
Targets jointly set by work group and supervision.
They divide up the necessary operations amongst themselves.
S4.
Line, circle or no group structure
Each worker makes the whole product.
Targets set by work group or by individual worker.
S5.
Square, rectangular or circular groups.
Each worker makes the whole product.
Targets set by work groups.
Group requisitions materials
Group trains new members.
Group does own maintenance.
Group liaises with Sales and Production Departments.

BABYCARE LTD

Technical solutions short-list

Give brief description

T3. Mechanically paced line with buffer stocks between workers.

T4. Mechanically paced line with workers able to alter the speed of the line to meet their fatigue level.

T5. Individual work stations with no mechanical pacing.

T1. Line as it is now mechanically paced.

Social solutions short-list

Give brief description

S2. Line-type work organisation with each worker assigned a number of operations.

S3. Line or circular work organisation with group given responsibility for allocating tasks amongst group members.

S4. Line, circle or individual work organisation with each worker doing a whole job.

S5. Circular work organisation with each individual doing a whole job. The group responsible for a number of ancilliary tasks.

S1 is the present form of work organisation. This is rejected as not meeting human objectives.

BABYCARE LTD

STEP 4 SET OUT POSSIBLE SOCIO-TECHNICAL SOLUTIONS.

It is now necessary to merge your short-lists of technical and social solutions.

The essential thing is to see which technical and social solutions are compatible with one another and to eliminate any technical or social solutions which cannot be fitted to a compatible social or technical solution.

Take each solution in turn and compare it with all the solutions on your other short-list. Where the two solutions could be operated together, mark this combination of solutions down as a possible socio-technical solution on the next page. It may be of course, that *all* your solutions are compatible with one another, in which case you can enter them all as possible socio-technical solutions.

Use *your* social and technical solutions, not ours, when you do this.

Do not rank the solutions in order of preference yet.

BABYCARE LTD

Possible socio-technical solutions short-list

Description	*Ranking*	*Description*	*Ranking*	*Description*	*Ranking*
1.		4.		7.	
2.		5.		8.	
3.		6.		9.	

BABYCARE LTD

STEP 5 RANKING SOCIO-TECHNICAL SOLUTIONS

The list of socio-technical solutions which *you* have just drawn up must now be ranked, before the most suitable system can be chosen. Turn back to Step 3, where you set out the efficiency and job-satisfaction advantages and disadvantages of the various solutions *you* put forward. Consider what you wrote on those pages and then try and rank the socio-technical solutions on the previous page from 1 to 9. If you find this rather difficult, you might like to award +1 to each advantage and −1 to each disadvantage until you achieve a score for each socio-technical solution, or to help differentiate further, you could award +2 to each major advantage and −2 to each major disadvantage.

When you feel that you have achieved a satisfactory ranking, enter this against each socio-technical solution on the previous page. It should now be clear which socio-technical solution you consider the best, but it is necessary to check that this solution is *completely* satisfactory before accepting it as the most suitable system.

Our ranked socio-technical solutions are shown on the next page.

BABYCARE LTD

Possible socio-technical solutions short-list

Description	Ranking	Description	Ranking	Description	Ranking
1. T3. Mechanically paced line with buffer stocks. + S2. Line-type work organisation. Each worker does a number of operations.	3	4. T4. Mechanically paced line with workers able to alter speed of line. + S3 Group is responsible for allocation of tasks.	8	7. T5. Individual work stations with no mechanical pacing. + S4. Line or circular type work organisation. Each individual doing a whole job. Targets set by group. Feedback to group.	1
2. T3. Mechanically paced line with buffer stocks. + S3. Group is responsible for allocation of tasks.	5	5. T5. Individual work stations with no mechanical pacing. + S2. Line type work organisation with each worker doing a number of operations.	2	8. T5. Individual work stations with no mechanical pacing. + S5. Circular work organisation. Each individual doing a whole job. Group responsible for ancilliary tasks.	6
3. T4. Mechanically paced line with workers able to alter the speed of line. + S2. Line-type organisation. Each worker does number of operations.	7	6. T5. Individual work stations with no mechanical pacing. + S3. Group given responsibility for allocating tasks.	4	9. T1. Line as it is now. + S2. Each worker assigned a number of operations.	9

BABYCARE LTD

CHECK YOUR CHOSEN SOCIO-TECHNICAL SOLUTION

Although you have already checked the technical and social solutions *separately*, it is now necessary to make sure that your combined socio-technical solution still meets the criteria laid down earlier. Turn back to Steps 1 and 2 and check that your chosen socio-technical solution meets the conditions set out there.

1. Does this solution meet *both* technical requirements and human needs?
2. Are sufficient resources available to achieve *both* the technical and social aspects of your chosen solution?
3. Do any of the constraints set out earlier make your chosen solution impossible?
4. Does this solution meet both the technical and social objectives which were set out in Step 2?

If you are satisfied that your chosen socio-technical solution is still viable, proceed to **Step 6** and prepare a detailed work design. If you find that your chosen solution has not met the criteria in Step 5 return to your short-list of socio-technical solutions and take the solution ranked next and go through **Step 6** again. Carry on until you find a socio-technical solution which meets the points listed above.

BABYCARE LTD

STEP 6 PREPARE A DETAILED WORK DESIGN FROM YOUR CHOSEN SOCIO-TECHNICAL SOLUTION

Work organisation and task-group structure

Make a drawing below of your reorganised department. Show the following:

1. The new arrangement of machines and work stations.
2. The flow of work through the department.
3. The task group structure. Indicate which individual work stations are part of larger task groups. If any work stations are isolated and do not interact with others, place a dotted circle around these.

BABYCARE LTD

Individual and group task structures

1. Set out below a list of all the tasks which workers will have to do if your socio-technical solution is implemented.

   List of tasks

2. Take either a representative task group and set out below, the number and kind of tasks the workers in it will have to complete *or* if workers are not sharing tasks take an individual work station and do the same.

3. Check the following:

	Yes	To some To some extent	No
(a) Do these tasks provide interest and variety and require skill?	☐	☐	☐
(b) Can the workers identify clearly the targets they much achieve?	☐	☐	☐
(c) Are there good feedback mechanisms to inform the workers of their performance?	☐	☐	☐
(d) Are there clear boundaries between the jobs of one task group or work station and another, so that the workers have a feeling of identity with their jobs?	☐	☐	☐
(e) Is the cycle time of the different tasks long enough to avoid a feeling of repetitive work but short enough to allow the workers to feel they are making progress with their work?	☐	☐	☐

BABYCARE LTD

Evaluation of solution

If you consider that the jobs you have created are as satisfying as they could be, while still achieving the technical objectives of the system, then you may accept this socio-technical solution as your final solution. If you have any doubts about the jobs which your chosen solution will create, go back to your short-list of socio-technical solutions, and take the solution which ranks next and proceed with Steps 5 and 6 again, until you are satisfied with your solution.

Set out below your arguments for selecting this particular solution:

BABYCARE LTD

Personnel implications of your socio-technical solution

What changes do you need to make in the following personnel activities to reinforce your socio-technical solution?

1. Recruitment.

2. Training.

3. Methods of wage payment.

4. Promotion policies.

5. Anything else (please indicate what this is).

BABYCARE LTD

Strategy for acceptance

Set out a strategy to convince the Factory Manager that your socio-technical solution is viable in terms of his objectives.

A new Food Packing Department

What changes would you make to your proposed socio-technical solution if you were designing an entirely new Food Packing Department?

BABYCARE LTD

TO COMPLETE THIS EXERCISE WE SET OUT OUR SOLUTION ON THE NEXT PAGES, TOGETHER WITH OUR ARGUMENTS FOR CHOOSING IT, A BRIEF NOTE ON ITS PERSONNEL IMPLICATIONS AND A STRATEGY FOR ACCEPTANCE.

IF YOUR SOLUTION DEVIATES GREATLY FROM OURS, IDENTIFY THE FACTORS IN YOUR ANALYSIS WHICH HAVE LED YOU TO TAKE A DIFFERENT ROUTE.

ONCE AGAIN WE STRESS THAT THERE IS NO RIGHT ANSWER. THE ANSWER THAT IS ARRIVED AT IS A CONSEQUENCE OF THE ANALYSIS THAT PRECEDES IT AND THE HUMAN OBJECTIVES THAT HAVE BEEN SET.

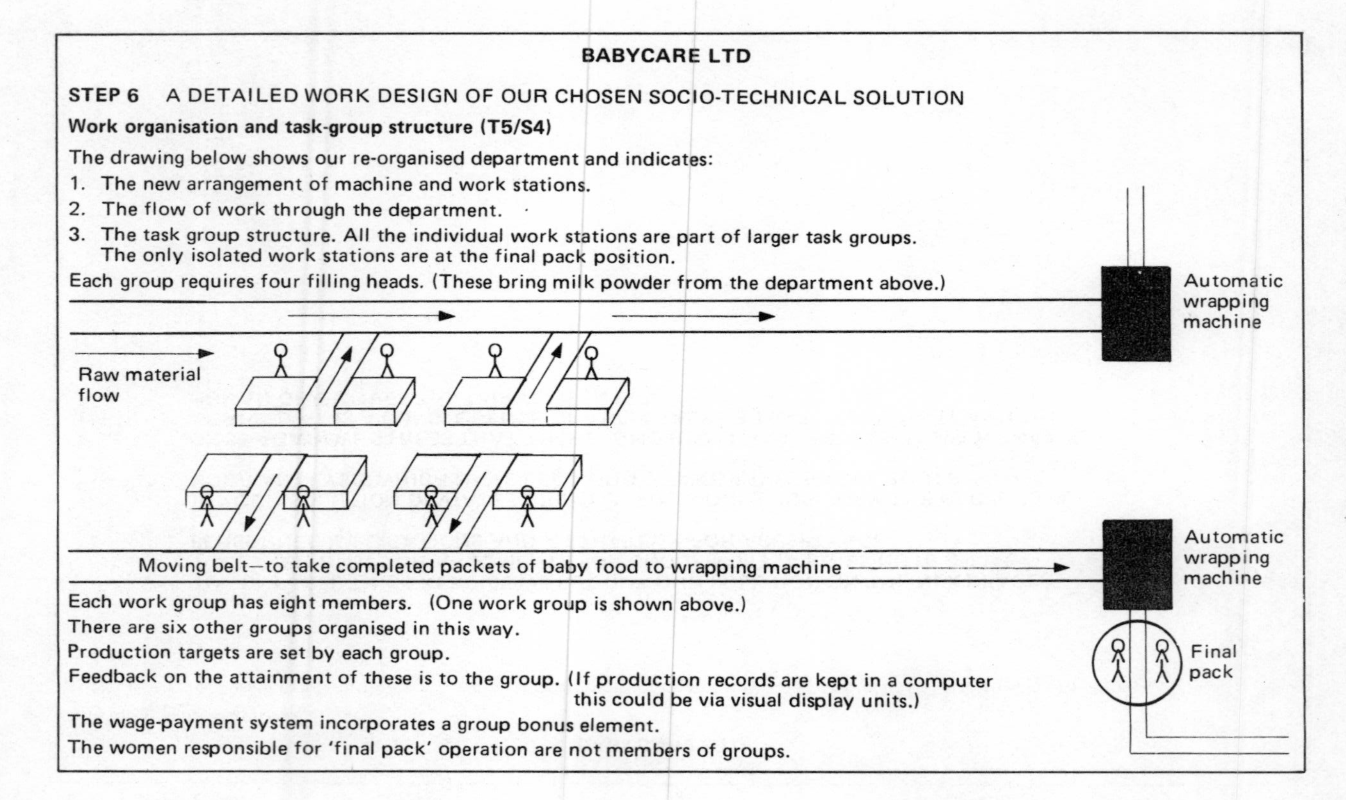

BABYCARE LTD

STEP 6 A DETAILED WORK DESIGN OF OUR CHOSEN SOCIO-TECHNICAL SOLUTION

Work organisation and task-group structure (T5/S4)

The drawing below shows our re-organised department and indicates:

1. The new arrangement of machine and work stations.
2. The flow of work through the department.
3. The task group structure. All the individual work stations are part of larger task groups. The only isolated work stations are at the final pack position.

Each group requires four filling heads. (These bring milk powder from the department above.)

Each work group has eight members. (One work group is shown above.)

There are six other groups organised in this way.

Production targets are set by each group.

Feedback on the attainment of these is to the group. (If production records are kept in a computer this could be via visual display units.)

The wage-payment system incorporates a group bonus element.

The women responsible for 'final pack' operation are not members of groups.

BABYCARE LTD

Individual task structures

1. Set out below are a list of all the tasks which workers will have to do if our socio-technical solution is implemented.

   List of tasks

   1. Fill bag with milk powder.
   2. Form cartons.
   3. Insert ticket and measure.
   4. Prepare bag for heat sealing.
   5. Insert bag into carton.
   6. Gum cartons.
   7. Final pack.
   8. Set output targets.

2. Each task group would take responsibility for the following activities:

   1. Fill bag with milk powder.
   2. Form carton.
   3. Insert ticket and measure.
   4. Prepare bag for heat sealing.
   5. Insert bag into carton.
   6. Gum carton.
   7. Set output targets.
   8. Monitor output targets.

3. We have checked the following:

	Yes	To some extent	No
(a) Do these tasks provide interest and variety and require skill?		√	
(b) Can the worker identify clearly the targets she must achieve?	√		
(c) Are there good feedback mechanisms to inform the worker of her performance?	√		
(d) Are there clear boundaries between the job of one worker and that of another, so that the worker has a feeling of identify with her job?	√		
(3) Is the cycle time of the different tasks long enough to avoid a feeling of repetitive work but short enough to allow the worker to feel she is making progress with her work?	√		

BABYCARE LTD

Evaluation of solution

Our arguments for selecting this particular solution are as follows:

This solution does not provide a high level of interest, variety and skill but it is thought that the existing labour force needs to become more stable before far-reaching changes are made. If tne women cope well with this form of work organisation then they can be given more responsibility. For example:

1. The group can be trained to do simple machine maintenance.
2. Each group takes responsibility for training its own new members.
3. Each group becomes responsible for the cleanliness of its own work area.
4. Groups can requisition necessary equipment from the stores.
5. Each group gets its own raw materials.
6. Production and Sales personnel representing group inputs and outputs, liaise directly with the work groups.

BABYCARE LTD

Personnel implications of our socio-technical solution

The changes we believe we should need to make in the following personnel activities to reinforce our socio-technical solutions are:

1. *Recruitment.* Efforts must be made to recruit intelligent workers, capable of assuming responsibility.
2. *Training.* The length and kind of training required by the new system must be carefully worked out. This must take account of the labour turnover rate, absenteeism and the learning capacity of the existing labour force. Task cycle times should not be longer than 10–12 minutes.
3. *Methods of wage payment.* A group bonus scheme may stimulate production. But this bonus should not form too large an element in the pay packet. 10% may be appropriate. The workers could be consulted on the wage-payment method they prefer.
4. *Promotion policies.* Promotion to a supervisory grade should be available for those who wish this.
5. *Group leaders.* Each group should have a section leader whose role is to ensure friendly, co-operative relationships within her group. She also has training and induction responsibilities for new group members.

BABYCARE LTD

Strategy for acceptance

Strategies that may convince the Factory Manager to try our socio-technical solution are as follows:

1. Make sure he is aware of the heavy financial costs of his existing production system.

2. Persuade him to take a course in work design and to visit other companies which have abandoned paced work systems.

3. Get approval of the proposed approach at board level. The Personnel Director can assist here.

4. Ensure that the workers themselves discuss and approve of the proposed work system and that they inform the Factory Manager of their approval.

A new Food Packing Department

What changes would we make to our proposed socio-technical solution if we were designing an entirely new Food Packing Department?

1. No changes to the solution but more space would assist work comfort. Independent work stations usually require more space than assembly lines and the existing department is already cramped.

2. Great attention needs paying to the material-handling part of the work system. Getting raw materials to the workers and removing finished products from them.

5 The ETHICS method — Exercise 2

MAIL-IT LTD, CUSTOMER ACCOUNTING SYSTEM

Background to the case-study exercise

This study concerns the main office of a mail-order firm called Mail-It Ltd.

Mail-It Ltd, is a firm operating an extensive mail-order service from its premises in Wolverhampton. The firm has been in this business for many years now and has built up a reputation for offering good-quality merchandise and providing a speedy service. It is experienced in the use of computers and has used a computer-based stock-control system for many years.

Over the last few years the business has been expanding rapidly and has become increasingly profitable. The main business objective of the firm is to continue this expansion for at least the next few years, by which time they forecast they will have reached capacity on their present site, and may need to consider opening new premises in a different part of the country.

However they feel their expansion plans may be constrained by two factors:

1. They are encountering difficulty in attracting sufficient staff to keep the present operation going at an acceptable level of efficiency.

2. Although their warehouse is large enough to cope with expansion, their office accommodation is becoming rather cramped and is creating difficulties for them in keeping those staff they manage to attract.

In this situation, they have started to explore the benefits of implementing a computerised system in their office, which will reduce their need for staff and require comparatively little space.

The basis of the mail-order operation is the agent, who acts on behalf of a group of customers in dealing with the firm. Every six months, the firm sends the agent a large, glossy catalogue showing the goods they have on offer. The agent shows the catalogue to her customers, who are usually friends and neighbours, and she encourages them to buy because the firm offers her a commission of 10p in the £ for all goods paid for. The agent takes orders from her customers and forwards them to the firm, who despatch the goods from their warehouse, either to the agent or directly to the customer. The goods are sold on a credit basis, which means that customers then have a fixed time to pay for the things they have received, at the rate of 5p in the £ each week for several months. The agent collects the weekly payments from each customer and sends them to the firm.

The task of the main office, with which we are concerned, is to deal with the accounting procedures necessary for the operation described above. They must debit the agent's account by the value of the goods her customers have ordered and credit the account as payments are received. Although these are basically simple operations, the sheer volume of transactions creates its own problems. There are 100000 agents altogether, most of whom send in a payment each week, which must be dealt with by the main office. This situation of a large volume of simple transactions is, of course, ideal for computerisation.

Despite the straightforward nature of the basic transactions, there are of course many other things to be done, which must also be built into the computer system as far as possible. For example: the order forms and payment slips which the agent sends in must be checked to see that they are correct; the order form must be checked to see that the size and colour of goods ordered are in fact available, that the price quoted is correct, and that there is a clear address to which to goods should be despatched. Once these things have been checked, a decision must be taken to ensure that the agent is credit worthy for the value of goods she is ordering. If there is any doubt, the order will be returned to her with an explanation. When credit worthiness has been proved, the agent's account is debited by the value of the goods, and a label is prepared to go on the package. The order form and label are then sent to the warehousemen, who select the goods from their shelves, pack and despatch them.

The procedure for dealing with the payments is, of course, much simpler. When the payment is received, the amount of money enclosed is checked against the amount stated on the payment slip. If these agree, the money is banked, and the payment slip is used as the firm's document to adjust the agent's account, that is, to reduce the agent's debit balance by the amount of the payment. A statement is sent to the agent each month, to inform her of the transactions on her account.

These two activities are the main ones in the office, but there are two other jobs which it is hoped the computer will also be able to tackle.

The first is to help with reminding agents who have fallen behind with their payments for goods. These agents are identified by checking through the records for all agents to see who has not sent in any payment for some time. Then a series of letters is sent reminding them that they should be paying weekly for the goods they have received. The letters start in a very pleasant tone, and become increasingly insistent, until eventually debt-recovery procedures may have to be put into operation.

The other area where a computer system may help is in dealing with the payment of commission to agents. The firm offers them 10p in the £ cash or 12½p in the £ in goods on all the money they have sent in to the firm, *not* on goods ordered. The firm relies on the agent to send in a commission claim, whenever she wishes to do so, but they must have some check that her claim agrees with their records of payments received.

A final area of activity which is very important, but on which the computer can be of only limited help, is in the queries and correspondence section. Queries may be received from agents about a wide range of topics, for example, complaining that goods ordered have not been received, or are not up to standard. Many queries are concerned with the accounting procedures, asking for explanations or drawing attention to errors which may have arisen. The image of the firm depends to a large extent on how well correspondence staff answer these complaints and queries.

At present the main office employs 500 staff, all of them

women, below the managerial grades. There are two main categories of women. About half are girls between the ages of 15 and 25, most of whom are filling the gap between leaving school and starting their families. The other main group are older women whose families have grown up. The majority of the women have no qualifications, and for those who have previously worked on the shopfloor in factories, these office jobs are a step up. The Mail-It office is situated in a residential area and so it is regarded as a convenient work place for local women, especially as it offers part-time hours for older women with families. The rates of pay are good for the area and overtime is unusual except at Christmas. However the rate of turnover is quite high — about 15 per month, mainly for domestic reasons, especially among the younger staff who almost all work full-time. The level of absenteeism is generally fairly low.

The main office is organised into five identical sections, so that each section keeps the records of about 20000 organisers. Each section has a manager, four supervisors and about 100 staff. The section does all the work associated with its own accounts, except for opening the envelopes and removing cash, which is done by a special section for security reasons. This special section sorts the documents for each of the five sections, and makes them up into batches of 10 documents; for example, 10 order forms or 10 payment slips, together with a batch summary form to show what is in the batch. The documents are batched in this way to make it easier to control the flow of documents round the section, and 10 documents are regarded as a convenient unit of work.

The jobs on the sections are all very simple and repetitive, with the cycle time for a batch of work being only a few minutes for most jobs. The overall impression is one of mass-production office work, with each girl doing a simple task on each batch, and passing it on to the next group of girls who perform a similar small procedure. Many of the jobs involve merely checking that the previous task has been completed correctly, and awareness that these checks exist tends to lead to faster but less accurate work. Very briefly, the jobs in the present office system, which the computer should be designed to take over are as follows:

1. *Scrutiny.* Checks that the goods ordered are offered in the catalogue in the colour and size requested and that the price on the order form is correct. (60)*
2. *Ledger-wells.* Takes out the firm's ledger card for the agent's account, so that the current transaction can be posted. (65)
3. *Credit sanction.* Checks that the agent is credit worthy for the goods she has ordered. (60)
4. *Balancing.* Brings the agent's account up to date if necessary, e.g., to amend for any goods returned or out of stock. (25)
5. *Pre-lister.* Adds up the value of orders and payments in the batch, to provide a check figure for the accounting machinist. (15)
6. *32 accounting machinist.* Posts the current transaction to the ledger card and arrives at a new balance figure for the account. (50)
7. *Checker.* Checks that the transaction has been correctly posted to the ledger card. (60)
8. *Separator.* Returns the ledger card to the ledger wells and the agent's documents go for filing. (15)

In addition, in each section there are: girls who file all letters and documents from each agent (35); some who also answer queries and write letters to agents (55); others who deal with their commission claims (35); and a further group who send reminder letters to bad payers (20).

The jobs are graded, with ledger-well clerks, separators and file girls at the lower end, rising to accounting machinists, correspondence clerks and commission clerks at the top end of the scale. The girls are paid a flat rate according to grade, and in addition, they earn a bonus on output, that is, on the number of batches of work they complete. This increases their wage by about one-third. The supervisors are also paid a

* The figures in brackets show the total number of staff doing the job before re-organisation.

flat rate and their bonus is dependent on the output of the girls under them.

In general, the supervision is helpful rather than dictatorial; though some supervisors tend to exercise strict discipline, possibly as a result of the payment system.

Trades-union activities were discouraged in Mail-It until recently. However, within the last few years they have recognised USDAW* for all supervisory and clerical staff. Membership is only around the 3 per cent level, and little interest is shown in union activities of any kind.

The management of Mail-It hope that a computerised system will enable them to reduce staff requirements by at least 200, as well as improve the efficiency of the office by increasing the speed of throughput of work, and reducing the number of errors. At the same time they hope to use this opportunity to improve the jobs which their staff do, as they feel this will be beneficial both in terms of job satisfaction for the staff, and in helping them to reduce turnover among the people they would like to keep, usually the women who prefer more demanding and interesting work.

They are now about to begin the design of a suitable computer system, and have decided to try and take both technical and social objectives into account in the design process. Because they are interested in increasing employee participation in new company developments, they have set up a systems design team which consists of equal numbers of technical computer experts from management services, departmental managers who will use the new system and experienced clerks from future user departments.

Imagine that you are the design team of Mail-It Ltd., and follow through the ETHICS method as a guide in finding the best possible socio-technical solution for your firm.

*Union of Shop, Distributive and Allied Workers.

MAIL-IT LTD

STEP 1 REQUIRES YOU TO MAKE A DETAILED DIAGNOSIS OF JOB-SATISFACTION NEEDS AND TO COMPLETE PAGES 144-6.

MAIL-IT LTD

STEP 1 DIAGNOSIS

Specify technical requirements and human needs.

Technical requirements

Physical factors To produce a computerised system of work which reduces staff numbers required, handles cash and orders efficiently, reduces the possibility of error.

Rate factors The system must process orders fast enough for goods to be sent from the warehouses the day orders are received.

Control factors There must be checking mechanisms to ensure that the correct order gets to the warehouse and that goods are sent the same day.

Human needs

Human needs in Mail-It can be identified by examining the questionnaire data on the next six pages.

MAIL-IT LTD

Clerks' Questionnaire

This is a simplified version of the questionnaire that was used in Mail-It. The data is presented under our five job-satisfaction headings.

1. If more than two-thirds of respondents agree with a positive statement this is a desirable feature and should, if possible, be incorporated into the new computer-based system. Similarly, if more than two-thirds disagree with a negative statement (e.g. I do not like my present job) this is a desirable feature of the work.

2. If more than two-thirds of respondents disagree with a positive statement (e.g. I find my work extremely interesting) this is an undesirable feature and a check must be made that it is not incorporated into the new system. Similarly, if more than two-thirds agree with a negative statement this is an undesirable feature of work.

The questionnaire shows the 'fit' between the work needs of clerks and the extent to which these needs are being met under the present pre-computer system. It also shows to some extent how clerks would ideally like their work needs to be met.

*indicates clerks' response.

MAIL-IT LTD

Clerks' Questionnaire

The knowledge 'fit'

	More than two-thirds agree	*More than two-thirds disagree*	*Answers falling between these limits*
1. My skills and knowledge are fully used in my present job.		*	
2. I find my work extremely interesting.		*	
3. I do not like my present job.			*
4. I should like a different job in Mail-It.			*
5. I should like to be doing a more difficult job.	*		
6. Mail-It does not provide enough opportunities for me to learn new things and develop my talents.	*		
7. I should like the opportunity of learning and doing a more challenging kind of job.	*		

MAIL-IT LTD

Clerks' Questionnaire

The psychological 'fit'

	More than two-thirds agree	*More than two-thirds disagree*	*Answers falling between these limits*
1. Status is important to me, I like to be respected.	*		
2. Working for Mail-It gives me a feeling of status.		*	
3. I carry considerable responsibility in my job.	*		
4. I should like to carry even more responsibility.			*
5. If I do good work, I feel management recognises this.		*	
6. I should like my work to receive more recognition.	*		
7. My job is a very secure one.		*	
8. I enjoy the opportunity for making friends which my job provides.	*		
9. There are not sufficient opportunities for promotion in this department.	*		
10. I should like to get promotion soon.			*
11. I want to achieve a great deal and get to the top		*	
12. I get my feelings of achievement from doing a good job.	*		

MAIL-IT LTD

Clerks' Questionnaire

The efficiency 'fit'

	More than two-thirds agree	*More than two-thirds disagree*	*Answers falling between these limits*
1. I could earn more money if I did not work for Mail-It.		*	
2. I am paid adequately for the work I do.	*		
3. I have to work too hard in my job.	*		
4. We are expected to be too accurate in our work.		*	
5. Methods of checking are too lax.	*		
6. Supervision here is strict.			*
7. I like to be left to get on with my work without interference from supervision.	*		
8. My supervisors give me all the help I need.	*		
9. The bonus system is fair to everyone.		*	
10. I get all the information and materials I require to do my job efficiently.			*

MAIL-IT LTD

Clerks' Questionnaire

The task-structure 'fit'

	More than two-thirds agree	More than two-thirds disagree	Answers falling between these limits
1. There is not much variety in my work.	*		
2. I should like a job which was less routine.	*		
3. There is little scope for me to use my own initiative.	*		
4. I should like a job which expected me to use my initiative more.	*		
5. In my job I have to rely on my own judgement and take decisions.			*
6. It would be better for me if I could take more decisions without having to ask someone first.			*
7. There is a lot of pressure in my job.	*		
8. I should like less pressure in work.	*		
9. I have clear-cut work targets and know when I achieve these.	*		
10. I should like clearer targets to aim at.			*
11. How well I can do my work depends on my co-workers.	*		
12. It would be better for me if I could do my work without depending on others.			*
13. I can organise and carry out my work the way I want.		*	
14. I should prefer more freedom to plan my work myself.	*		
15. I believe my job to be important and others recognise its importance.	*		
16. I should prefer a job which made a larger and more important contribution to the work of the department.	*		

MAIL-IT LTD

Clerks' Questionnaire

The ethical 'fit'

	More than two-thirds agree	*More than two-thirds disagree*	*Answers falling between these limits*
1. Top management in Mail-It is too ruthless.		*	
2. Managers and workers in Mail-It are very friendly.	*		
3. Mail-It puts production above the interests of its employees.			*
4. Mail-It looks after the welfare of its staff very well.	*		
5. A person's character and experience count for more here than formal qualifications.			*
6. I believe that character and experience are more important than qualifications.	*		
7. Communication in this department is very poor.		*	
8. In my section, we are told as much as we want to know about day-to-day matters.	*		
9. Mail-It senior management is out of touch with the way the clerks feel.	*		
10. We have sufficient say in the way Mail-It is run.			*
11. Clerks should be consulted more about major decisions that are to be made.	*		
12. We cannot sufficiently influence decisions about changes which are made in the way we do our work.	*		

(Asterisk or make written comments as appropriate)

MAIL-IT LTD

Analysis of Human Needs

	Satisfactory aspects	*Should be incorporated into new system?*	*Unsatisfactory aspects*	*Must be improved in new system*	*Suggestions on how this can be achieved*
Knowledge 'fit'					
Use of knowledge					
Self-development					
Psychological 'fit'					
Status					
Responsibility					
Recognition					
Job security					
Social relationships					
Promotion					
Achievement					

MAIL-IT LTD
Analysis of Human Needs

	Satisfactory aspects	*Should be incorporated into new system?*	*Unsatisfactory aspects*	*Must be improved in new system*	*Suggestions on how this can be achieved*
Efficiency 'fit'					
Salary					
Amount of work					
Standards					
Controls					
Supervision					
Task-structure 'fit'					
Work variety					
Initiative					
Judgement and decisions					
Pressure					
Targets					
Dependency on others					
Autonomy					
Task identity					

MAIL-IT LTD

Analysis of Human Needs

	Satisfactory aspects	*Should be incorporated into new system ?*	*Unsatisfactory aspects*	*Must be improved in new system*	*Suggestions on how this can be achieved*
Ethical 'fit'					
Company ethos					
Communication					
Consultation					

Please summarise your analysis and enter your summary into the box below, noting the most important areas that require improvement.

Summary of required improvements	

MAIL-IT LTD

OUR ANALYSIS IS SHOWN ON THE NEXT THREE PAGES. IF YOURS IS DIFFERENT FROM OURS, IDENTIFY THE REASONS WHY. REMEMBER YOUR ANALYSIS MAY BE BETTER THAN OURS.

MAIL-IT LTD

Analysis of Human Needs

(Asterisk or make written comments as appropriate)

	Satisfactory aspects	*Should be incorporated into new system?*	*Unsatisfactory aspects*	*Must be improved in new system*	*Suggestions on how this can be achieved*
Knowledge 'fit'					
Use of knowledge			Skills and knowledge under-utilised. Work not interesting. Work not difficult enough.	* * *	Creation of larger jobs requiring greater use of different skills.
Self-development			Not enough opportunities for self-development.	*	Job progression from simple to increasingly difficult jobs.
Psychological 'fit'					
Status			Status wanted but not provided.	*	Consider badges of status or status-conferring work environment.
Responsibility	Has responsibility.	Yes.			Try and provide same amount of responsibility.
Recognition			Management not seen as recognising good work.	*	Through management education programme associated with new system.
Job security			Job security seen as threatened by new system.	*	'No redundancy' policy.
Social relationships		Yes.			Ensure new system provides opportunities for social relationships.
Promotion			Promotion opportunities seen as inadequate.	*	Associate new personnel policies on promotion with computer system. Clerks are prepared to wait for this. Check system does not inhibit promotion opportunities.
Achievement	Clerks feel achievement if they do their work well.	Yes.		*	This should be maintained if more interesting jobs are provided.

MAIL-IT LTD

Analysis of Human Needs

	Satisfactory aspects	*Should be incorporated into new system?*	*Unsatisfactory aspects*	*Must be improved in new system*	*Suggestions on how this can be achieved*
Efficiency 'fit					
Salary	Level of pay.	Yes.	The existing bonus scheme.	*	Check bonus is appropriate for new system. If not, alter it. Check existing salary levels and grades are appropriate for new system. If not, alter them.
Amount of work			Clerks feel overloaded.	*	Ensure workload is not increased. Feelings of overload may be due to low interest.
Standards	*		Not seen as good enough.	*	Note existing level of accuracy is approved.
Controls			Not seen as good enough.		See new system ensures fewer missed errors.
Supervision	*	Yes.			Ensure supervisory control stays at same level
Task-structure 'fit'					
Work variety			Not enough variety in the work.	*	Larger jobs involving the use of more skills.
Initiative			Not enough scope for initiative.	*	Jobs which incorporate some uncertainty so that choices have to be made.
Judgement and decisions	No strong feelings.		No strong feelings.		
Pressure			Too much pressure.	*	Better distribution of workloads throughout day will reduce pressure.
Targets	Satisfaction with targets.				Ensure that targets and feedback are incorporated into new system.

MAIL-IT LTD
Analysis of Human Needs

	Satisfactory aspects	*Should be incorporated into new system?*	*Unsatisfactory aspects*	*Must be improved in new system*	*Suggestions on how this can be achieved.*
Task-structure 'fit' (continued)					
Dependency on others	Clerks are dependent on the accuracy of others but there are no strong feelings about this.		See satisfactory aspects.		
Autonomy			More autonomy is required.	*	Larger jobs with well defined beginnings and ends and few constraints on how work is done.
Task identity	Jobs seen as important.	Yes.	More important jobs wanted.	*	Again, through larger, more meaningful sets of tasks.
Ethical 'fit'					
Company ethos	Mail-It is perceived as a benevolent company that values experience in its staff.				Ensure that nothing happens to change clerks' attitudes to the company.
Communication	*	Yes.			Ensure that communication within Mail-It continues to be good.
Consultation			Better consultation required on major changes such as the new computer system.		Establish improved consultation mechanisms to cope with change.
Summary of required improvements	Knowledge fit—Better use of skills and knowledge. Opportunities for self-development. Psychological fit—Want status, recognition, job-security assurances, better promotion opportunities. Efficiency fit—Better work controls, reduce overload of work, amend bonus system. Task-structure fit—More variety, scope for initiative, less pressure, more autonomy, more task identity. Ethical fit—Better communication in Mail-It. More consultation on 'change'.				

MAIL-IT LTD

STEP 2 IDENTIFIES THE CONSTRAINTS WITHIN WHICH YOU MUST WORK WHEN DESIGNING THE NEW SYSTEM. IT ALSO TELLS YOU THE RESOURCES THAT ARE AVAILABLE TO HELP YOU DESIGN THE TECHNICAL AND SOCIAL PARTS OF THE SYSTEM.

THE TECHNICAL (EFFICIENCY) OBJECTIVES WHICH YOUR SYSTEM MUST MEET ARE GIVEN. YOU MUST SET THE HUMAN OBJECTIVES WHICH WILL ENABLE YOU TO IMPROVE JOB SATISFACTION. THESE WILL BE DERIVED FROM YOUR ANALYSIS OF HUMAN NEEDS.

MAIL-IT LTD

STEP 2 SOCIO-TECHNICAL SYSTEMS DESIGN

Identify constraints

Identify technical and business constraints on the design of the system

The system must be sufficiently robust to reduce the amount of breakdown time to the minimum. There should also be adequate backup and recovery procedures, to maintain a high level of service to the customer.

The hardware must be used as economically as possible, that is, the maximum possible use must be made of the hardware which is compatible with the social constraints and social objectives.

Identify social constraints on the design of the system

The overall number of staff must be reduced (by about 200 if possible), but there must be no redundancy.

The jobs created must be appropriate for the skill levels of the existing staff, bearing in mind that they feel their skills are under-utilised at present.

MAIL-IT LTD

Identify resources

Identify resources available for the technical system

The existing Management Services Department has experienced systems analysts and programmers available, though more will need to be recruited.

Assistance may also be available from the computer manufacturer supplying the hardware.

The board has allocated sufficient funds for a major system and is prepared to leave the hardware decision to the experts, so long as the budget is not exceeded.

Identify resources available for the social system

Personnel and training facilities are available already and two extra full-time senior people can be recruited for planning and supervision of the training programme as well as one additional person for specialised training.

Assistance is also available from the Organisation and Methods Department.

Funds are available for the recruitment of additional temporary staff, should this prove necessary during parallel running and implementation.

MAIL-IT LTD

Specify broad technical and human objectives

Technical (efficiency) objectives

To introduce a computer system to deal with customer orders and accounts. This system should:

1. Reduce staff requirements which are becoming increasingly difficult to meet because of a shrinkage in the labour market.
2. Improve the level of service to customers.

Human (job-satisfaction) objectives

These should be derived from your analysis of human needs.

CHECK THAT YOUR TECHNICAL AND HUMAN OBJECTIVES ARE COMPATIBLE

MAIL-IT LTD

THE HUMAN OBJECTIVES WHICH WE HAVE SET ARE

Human (job-satisfaction) objectives

The creation of a work-force which is efficient, motivated and identified with Mail-It interests and which has job security, job satisfaction and opportunities for personal-growth development.

MAIL-IT LTD

STEP 3 REQUIRES YOU TO SET OUT A NUMBER OF ALTERNATIVE TECHNICAL SOLUTIONS AND TO SPECIFY THE EFFICIENCY AND JOB-SATISFACTION ADVANTAGES AND DISADVANTAGES OF EACH OF THESE.

BY TECHNICAL ALTERNATIVE WE MEAN HERE THE DIFFERENT WAYS IN WHICH THE COMPUTER CAN FORM PART OF THE WORK SYSTEM. FOR EXAMPLE, YOU WILL NEED TO CONSIDER DIFFERENT WAYS OF HANDLING DATA INPUT AND OUTPUT AND THE TECHNOLOGY THAT CAN BE USED FOR THIS. AS THE MAIL-IT COMPUTER IS BEING USED TO UPDATE CUSTOMER RECORDS, YOU WILL NEED TO CONSIDER WHETHER YOU WISH TO HAVE A BATCH SYSTEM, AN ON-LINE SYSTEM THAT WORKS IN BATCH MODE OR AN ON-LINE/REAL-TIME SYSTEM.

(THOSE WHO ARE UNFAMILIAR WITH COMPUTER SYSTEMS WILL FIND INFORMATION ON THE DIFFERENT OPTIONS IN APPENDIX B).

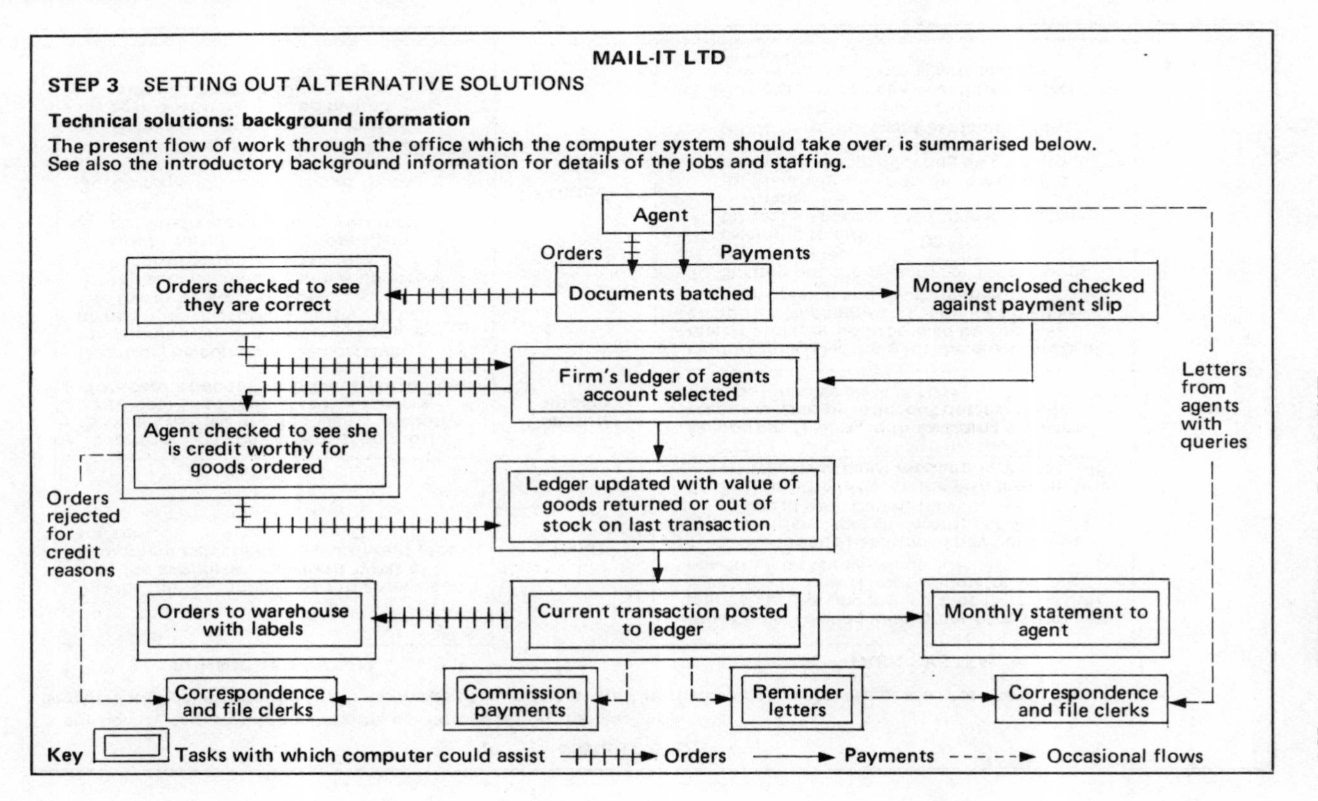
MAIL-IT LTD
STEP 3 SETTING OUT ALTERNATIVE SOLUTIONS
Technical solutions: background information
The present flow of work through the office which the computer system should take over, is summarised below.
See also the introductory background information for details of the jobs and staffing.
Agent
Orders
Payments
Orders checked to see they are correct
Documents batched
Money enclosed checked against payment slip
Firm's ledger of agents account selected
Letters from agents with queries
Agent checked to see she is credit worthy for goods ordered
Ledger updated with value of goods returned or out of stock on last transaction
Orders rejected for credit reasons
Orders to warehouse with labels
Current transaction posted to ledger
Monthly statement to agent
Correspondence and file clerks
Commission payments
Reminder letters
Correspondence and file clerks
Key
Tasks with which computer could assist
Orders
Payments
Occasional flows

MAIL-IT LTD

Your new system should include the operations set out below.
(Figures in brackets are the firm's estimates of numbers of staff needed to deal with various parts of the system.)

TECHNICAL SYSTEM		TASK SYSTEM
Input. The equipment used to enter data into the computer—different types of equipment can create very different jobs for staff.	Input	*Input jobs.* Dealing with order forms and payment slips showing money sent from 10 000 agents each week. Great volume of simple repetitive tasks involving: 1. Checking and batching of payment slips and orders from the agent, and totalling values in each batch. (45) 2. Entering this information into the computer via whatever input medium is chosen. (45)
Computer. The computer can do most of the work needed to bring agents' accounts up-to-date—and then throw out errors. Work can be done either in batches or on-line.	Computer update	*Computer.* Dealing with updating of agents' accounts—and printing out details of rejects—does not involve clerical staff.
Output. The computer can produce information on to a terminal, paper printout, or microfilm. Output to be dealt with by staff will be: 1. Input and validate—shows up errors in information entered. (40) 2. Credit reject—shows those agents not credit worthy for goods they have ordered. (40) 3. Commission reject—shows those commission claims from agents with errors or miscalculations. (30) 4. Reminder letters—shows those agents who have fallen behind with payments. (20)	Output	*Output jobs.* Dealing with the 4 printouts of rejects and queries produced by the computer (see opposite). The jobs involve problem-solving and decision-making. 1. Sorting out the reasons for transactions being rejected. 2. Deciding to either: (a) Make a correction if an error has been made. (b) Manually override the computer's decision. e.g. allowing an order to go through for special reasons. (c) Write to the agent explaining why the rejection has occurred. 3. Correspondence clerks will be needed to answer queries from agents and also filing clerks. (75)

Flow of work through office ↓

MAIL-IT LTD

STEP 3 SETTING OUT ALTERNATIVE SOLUTIONS

Technical solutions

It may be easier to break the technical system down into 3 stages:

1. *Input* The method by which data gets into the computer.
2. *Updates* Whether accounts are processed straight away in real time, or later in batches—and how often the batches are run.
3. *Output* Whether the information produced by the computer is printed out at once on-line through a terminal, or printed out on paper or microfilm, and what form the printout should take.

Set out on the next three pages several different technical solutions which you think might be appropriate for the new computer system, bearing in mind the objectives which have been set. Four or five alternatives would be a reasonable number to consider, though of course you may consider more or fewer alternatives if you wish. Some alternatives might differ only in one aspect, e.g. method of input or the form of the printout, so it may help to group these together under a broad heading with the differences as subheadings. For each solution or group of solutions, set out the technical *and* social advantages and disadvantages which you think it may have.

MAIL-IT LTD

Description of technical solution	*Advantages*	*Disadvantages*
(a) *Input* (b) *Update* (c) *Output*	*Efficiency advantages*	*Efficiency disadvantages*
	Job-satisfaction advantages	*Job-satisfaction disadvantages*
(a) *Input* (b) *Update* (c) *Output*	*Efficiency advantages*	*Efficiency disadvantages*
	Job-satisfaction advantages	*Job-satisfaction disadvantages*

MAIL-IT LTD

Description of technical solution	*Advantages*	*Disadvantages*
(a) *Input*	*Efficiency advantages*	*Efficiency disadvantages*
(b) *Update*		
(c) *Output*	*Job-satisfaction advantages*	*Job-satisfaction disadvantages*
(a) *Input*	*Efficiency advantages*	*Efficiency disadvantages*
(b) *Update*		
(c) *Output*	*Job-satisfaction advantages*	*Job-satisfaction disadvantages*

MAIL-IT LTD

Description of technical solution	*Advantages*	*Disadvantages*
(a) *Input*	*Efficiency advantages*	*Efficiency disadvantages*
(b) *Update*		
(c) *Output*	*Job-satisfaction advantages*	*Job-satisfaction disadvantages*
(a) *Input*	*Efficiency advantages*	*Efficiency disadvantages*
(b) *Update*		
(c) *Output*	*Job-satisfaction advantages*	*Job-satisfaction disadvantages*

MAIL-IT LTD

Now do a preliminary evaluation of the alternative *technical* solutions you have put forward.

For each solution ask yourself:

1. Does it achieve the technical requirements which we specified in Step 1?
2. Is it limited in any way by the constraints identified earlier?
3. Are the available resources which were identified earlier adequate to achieve this solution?
4. Does it meet the objectives which were set out earlier?

You can now draw up a short-list of technical solutions which still seem reasonable after they have been examined in this way.

Try and keep your short-list down to three or four solutions. Write out your short-list on the next page.

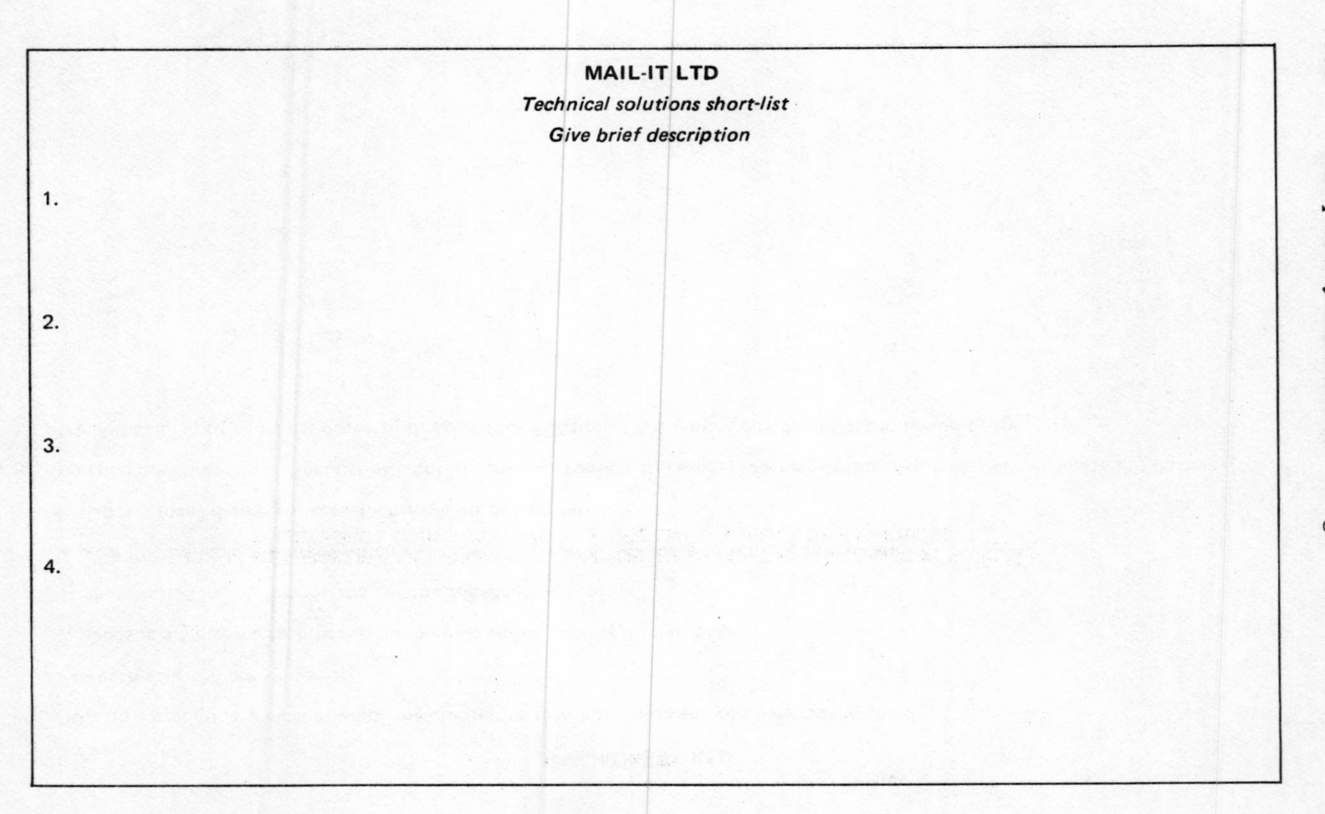

MAIL-IT LTD

Technical solutions short-list

Give brief description

1.

2.

3.

4.

MAIL-IT LTD

REMEMBER THERE ARE NO COMPLETELY RIGHT OR WRONG TECHNICAL SOLUTIONS. OUR SUGGESTIONS ARE ON THE FOLLOWING PAGES.

MAIL-IT LTD
Technical Solution

Description of technical solution	*Advantages*	*Disadvantages*
T1. (a) *Input* Key to tape—the key punch operator keys information straight on to magnetic tape. (b) *Update* Batch processing from magnetic tape. (c) *Output* 1. Paper printout for rejects. 2. List of orders and labels to warehouse. 3. Statement to agent.	*Efficiency advantages* 1. Cuts out one stage in system, i.e. transcript to magnetic tape. 2. Cheapest and easiest method to update when enough information is ready. 3. Easy to recover in case of breakdown *Job-satisfaction advantages* 1. Similar to using typewriter, so it is easier for girls who have been typists. 2. Paper printout is fairly easy to read, if it is well set out.	*Efficiency disadvantages* 1. Needs 12 weeks training, therefore very dependent on operators which could be difficult, e.g. in flu epidemic. 2. Time delay before rejects are dealt with because batch processing system holds orders up. 3. Paper printout tends to be bulky. *Job-satisfaction disadvantages* 1. Girls do not see what they are punching, can be frustrating. 2. Batch processing system can cause problems for staff dealing with rejects—they are delayed if computer breaks down and then have to cope with build-up of work. 3. Paper printout is heavy to handle and store.
T2. (a) *Input* Document reader, which automatically scans specially prepared documents. (b) *Update* As in 1 above. Batch processing—several times a day, when sufficient input has been read in. (c) *Output* As in 1 above. Paper printout of rejects. List of orders and labels for warehouse. Monthly statements for agents.	*Efficiency advantages* 1. Pre-printed documents could be sent to agent and fed straight into reader on return. Considerable cost savings. 2. Reduces number of documents needed to run the system. Cost savings. 3. Reduces chances of error. *Job-satisfaction advantages* 1. Reduces number of staff needed on input jobs, which tend to be boring and repetitive.	*Efficiency disadvantages* 1. Needs high level of competence on the part of the agent to fill in documents. 2. Doubtful if any document reader is robust enough to do the job efficiently. 3. Utterly dependent on reader for all input—could be high risk. *Job-satisfaction disadvantages* 1. Some people would prefer a boring job to none at all.

MAIL-IT LTD
Technical Solution

Description of technical solution	*Advantages*	*Disadvantages*
T3. (a) *Input* Paper tape — girls punch information on to paper tape, and it is then transferred to magnetic tape. (b) *Update* As in 1 above. Batch processing from magnetic tape. (c) *Output* As in 1 above. Paper printout for rejects. List of orders and labels for warehouse. Monthly statements for agents.	*Efficiency advantages* 1. Reliable and cheap method. 2. Best method of dealing with some documents e.g. summaries of items in each batch.	*Efficiency disadvantages* 1. Inflexible method — involves considerable time delay before rejects can be corrected. 2. Needs experienced operators, therefore dependent on them
	Job-satisfaction advantages 1. Uses typewriter keyboard, so girls familiar with that can use punch machines fairly quickly.	*Job-satisfaction disadvantages* 1. Boring, monotonous job which needs a high level of concentration. 2. It is difficult for the girls to read the tape they have punched.
T4. (a) *Input* Visual display units on-line to small computer — which does some checking and creates a magnetic tape. (b) *Update* As in 1 above. Batch processing from magnetic tape. (c) *Output* As in 1 above. Paper printout for all output.	*Efficiency advantages* 1. Flexible system — mistakes can be corrected immediately before they go on to magnetic tape. 2. Fast system — orders can be credit checked at once and then go straight to warehouse. 3. Needs only 5 weeks training, therefore not as dependent on girls.	*Efficiency disadvantages* 1. Need a larger computer, or an additional small computer to run the VDUs. 2. Need more expensive equipment to store agent's accounts in computer for easy access. 3. VDUs themselves are relatively costly.
	Job-satisfaction advantages 1. Reasonably satisfying job for the VDU operators, as they are interacting with the computer via VDUs. 2. Operator can understand what the computer displays on screen — no special languages used or holes in tape, etc.	*Job-satisfaction disadvantages* 1. Can still be a fairly repetitive job when dealing with a lot of documents which are all the same. 2. Can sometimes lead to eyestrain and headaches from looking at screen continuously.

MAIL-IT LTD
Technical Solution

Description of technical solution	*Advantages*	*Disadvantages*
T5. (a) *Input* As in 4 above. Visual display units on-line creating magnetic tape. (b) *Update* As in 1 above. Batch processing from magnetic tape. (c) *Output* Rejects on microfilm. List of orders and labels on paper. Paper statements for agents.	*Efficiency advantages* 1. Printout on microfilm takes up much less space for handling and storage than paper printout. This provides cost savings. *Job-satisfaction advantages* 1. Less bulky for girls to handle and store.	*Efficiency disadvantages* 1. Extra equipment for printing out and reading is required. 2. May need special office conditions e.g. darkened area. *Job-satisfaction disadvantages* 1. Can be more difficult to find rejects and make record of action. 2. May be less pleasant working conditions and eyestrain.
T6. (a) *Input* Via terminal, on-line direct to computer. (b) *Update* Real time — agent's account updated at once. (c) *Output* Rejects via terminal on-line direct from computer. Orders and labels direct to warehouse via terminal from computer. Paper statements for agents.	*Efficiency advantages* 1. Very fast system, including correction of mistakes and accounts always up to date. 2. Orders get to warehouse more quickly for quicker despatch to customer. *Job-satisfaction advantages* 1. Can be satisfying to interact directly with the computer — raising queries and receiving replies. 2. Girls get information on accounts straight away — and also information on their own performance.	*Efficiency disadvantages* 1. Advanced system — expensive to develop. 2. Need more expensive equipment to update immediately. 3. More chance of breakdowns. 4. Agent's accounts do not really need to be always up to date as it is a credit system. *Job-satisfaction disadvantages* 1. Most girls would be operating terminals therefore the range of jobs available for staff is reduced, especially for those who do not want to operate a machine.

MAIL-IT LTD

Technical solutions short-list

We decided that the document reader was too unreliable to meet the constraints imposed by Mail-It, so solution T2 was excluded from the short-list.

Solution T6 was also excluded from our short-list, because it was too expensive and sophisticated for the job Mail-It wanted doing. Accounts do not need to be absolutely up to date since it is a credit system. Our *technical solution short-list* was as follows:

T1. Key-to-tape input. Batch-processing update. Paper printout.

T3. Paper-tape input. Batch-processing update. Paper printout.

T4. Visual display units on-line for input. Batch-processing update. Paper printout.

T5. Visual display units on-line for input. Batch-processing update. Microfilm output for rejects.

MAIL-IT LTD

STEP 3 SOCIAL SOLUTIONS

Background information for social solutions

You have already examined the 'fit' between employee needs and experience on our five variables; (1) Use of skills and knowledge; (2) Psychological needs; (3) The methods used to ensure efficiency, including salary, work controls and support services; (4) The need for a task structure which meets needs for variety, autonomy etc.; (5) The need to work for a firm whose ethics and values fit those of the employee. The firm too has requirements on all these variables and these influence its policies. Mail-It's policies are set out below:

1. *Knowledge policy*

   Mail-It recruits people with a low level of skill, mainly from the local secondary-modern schools or local factories. They find difficulty in recruiting women with a higher level of skill, because the firm has a reputation of being a 'white-collar' factory. Most jobs are very simple and sufficient knowledge can be gained to do them in 2–3 weeks. There is a clear demarcation between jobs, so that each person only knows about her own job. Because of this and the low skill levels there are a great many controls included in the system.

2. *Policies for meeting employees' psychological needs*

   The management of the firm tends to be paternal, and has tried to create a family feeling. For example, everyone refers to the Chairman of the company by his christian name. Because the firm is labour intensive they need a high level of commitment on the part of their employees, especially at peaks like the Christmas rush. They also employ a fair proportion of part-timers and they provide a wide variety of working hours in order to attract them. Many would not be able to find suitable employment in the district, if it were not for these special arrangements.

MAIL-IT LTD

STEP 3

Background information for social solutions (continued)

3. *Policies for ensuring efficiency*

The standard of work is fairly low because the firm seems to demand 'quantity not quality'. This policy is apparent in the incentive bonus scheme which is earned individually and forms about 33% of take-home pay. The bonus in the supervisory grades is dependent on the output of those under them and this results in a great deal of pressure being put on the girls to 'get the work through'. The firm has instituted lots of checks into the system, but the knowledge that these exist only seems to encourage poor quality work.

The firm places a great deal of importance on a high level of performance, as shown by the bonus system, which provides a yardstick for each individual's performance. There are also 'merit rises', for which everyone is rated every 6 months, on performance and efficiency etc. They are very strict about clocking on and off and everyone from the juniors to the Chairman has a clock card.

4. *Policy on task structure*

The work of the office is sub-divided into many simple tasks, with the documents being passed from hand to hand along the line. The firm requires people who can keep at a repetitive job and achieve a good output. The jobs require very little skill and initiative.

5. *Ethical approach*

The firm has a reputation for ruthlessness amongst its younger staff, especially towards those who do not achieve its performance standards. Older staff see it as paternal and welfare minded and do not question the bureaucratic system. Mail-It tries hard to communicate with its staff but it is not very successful. There is a system of joint consultation but this is largely ineffective. The firm has discouraged union membership for many years but has recently recognised a non-militant union and is encouraging its staff to join.

MAIL-IT LTD

STEP 3 SETTING OUT ALTERNATIVE SOCIAL SOLUTIONS

Now proceed to set out alternative social solutions, though at this stage you may think about them independently of the technical solutions which you have just set out. Again, you may find it easiest to think in terms of the input and output tasks which need to be performed, but of course the update tasks are performed by the computer itself. It may also be helpful to think of the social system as a two-part system. The first part is the organisation of the total workflow into its various departments or sections. Later on comes the design of the actual work which is undertaken in each of these departments. At present, concentrate on the departmental organisation, and consider the more detailed job design in Step 6.

MAIL-IT LTD
Social Solution

Description of social solution	*Advantages*	*Disadvantages*
(a) *Input*	*Job-satisfaction advantages*	*Job-satisfaction disadvantages*
(b) *Output*	*Efficiency advantages*	*Efficiency disadvantages*
(a) *Input*	*Job-satisfaction advantages*	*Job-satisfaction disadvantages*
(b) *Output*	*Efficiency advantages*	*Efficiency disadvantages*

MAIL-IT LTD
Social Solution

Description of social solution	*Advantages*	*Disadvantages*
(a) *Input*	*Job-satisfaction advantages*	*Job-satisfaction disadvantages*
(b) *Output*	*Efficiency advantages*	*Efficiency disadvantages*
(a) *Input*	*Job-satisfaction advantages*	*Job-satisfaction disadvantages*
(b) *Output*	*Efficiency advantages*	*Efficiency disadvantages*

MAIL-IT LTD

When you have set up several alternative ways of organising the departments concerned, do a preliminary evaluation of your social solutions as you did for the technical solutions as follows:

1. Do the solutions meet the social needs which you identified in Step 1?
2. Are any of the solutions constrained by the conditions which you identified earlier?
3. Have you got the resources available to achieve each of the solutions you have set up?
4. Do all the solutions achieve the social objectives which you established earlier?

Having checked all your social solutions against these four criteria, eliminate any about which you are doubtful, and draw up a short-list of the remainder; if possible no more than three or four solutions. Enter your short-list of social solutions on the right-hand side of the next page. Copy your list of preferred technical solutions which you wrote in on page 169 onto the left-hand side.

MAIL-IT LTD.

Technical solutions short-list *Give brief description*	*Social solutions short-list* *Give brief description*
1.	1.
2.	2.
3.	3.
4.	4.

MAIL-IT LTD

OUR SUGGESTIONS ARE ON THE FOLLOWING PAGES.

MAIL-IT LTD
Social Solution

Description of social solution	*Advantages*	*Disadvantages*
S1. (a) *Input* The office is divided into groups with each girl doing *all* stages of the work.	*Job-satisfaction advantages* 1. Staff like working in groups. This creates feeling of group identity. 2. The work is interesting as each girl has a variety of tasks. 3. It is more satisfying to see the work through from beginning to end rather than to do just one small task. 4. Staff are able to get to know their own group of customers and this could lead to improved customer service.	*Job-satisfaction disadvantages* 1. These jobs would be of a much higher grade than the present jobs, and some staff might find them too difficult. 2. A much longer training period would be needed for these jobs and so it would increase the social cost for the firm.
(b) *Output* See above.	*Efficiency advantages* 1. This arrangement provides work flexibility. If some staff are ill, all the others know their work. 2. The quality of work should be better and need fewer checks if one person is seeing the job right through.	*Efficiency disadvantages* 1. It would be difficult to organise work in this way and still make full use of the *input* equipment. 2. This arrangement could have security problems so good controls will be needed. 3. The throughput of work may be rather slower if one person is doing all stages of the work.
S2. (a) *Input* The office is divided into groups which deal with *all* stages of the work, — but each girl is responsible for *one* task only, and passes her work on to the next girl in her group.	*Job-satisfaction advantages* 1. The staff will be working in a group, even though they are only doing one task, so they can identify with that group. 2. Staff will be dealing with a specific group of customers, so they will get to know them. 3. The jobs will be very similar to the type of jobs in the present system.	*Job-satisfaction disadvantages* 1. Some staff may find it boring and repetitive if they are only doing one task. This is no improvement on the present arrangement under the pre-computer system.
(b) *Output* See above.	*Efficiency advantages* 1. The work may go through more quickly if each person is concentrating on one task.	*Efficiency disadvantages* 1. This arrangement tends to be rather inflexible if some staff are ill or on holiday.

MAIL-IT LTD
Social Solution

Description of social solution	*Advantages*	*Disadvantages*
S3. (a) *Input* Each stage of the work is done by a group concentrating on nothing else — each girl deals with only one type of document.	*Job-satisfaction advantages* None.	*Job-satisfaction disadvantages* 1. This arrangement is really a mass-production clerical work system, with limited job interest for staff. 2. Because all members of the group are doing the same task, it is difficult for staff to understand the total office workflow.
(b) *Output* See above.	*Efficiency advantages* 1. It may be easier to control the flow of work, especially the input of data, if all the input machines are together.	*Efficiency disadvantages* 1. This arrangement tends to be inflexible especially if the computer breaks down. It would be difficult to move staff to other jobs.

MAIL-IT LTD

Social Solutions

The three social solutions set out above may be represented diagramatically for those who find that diagrams assist comprehension. Suppose the whole office were represented by a square, with the arrow showing the direction of the flow of work and with the computer in a central position, so that tasks done before the computer were broadly regarded as 'input tasks' — and those dealing with items produced by the computer as 'output tasks', as follows:

Agent

Input tasks

COMPUTER

Output tasks

The arrangement of the departments dealing with input and output tasks can then be indicated and the work done by *one* person shown as a shaded area. The social solutions set out can now be represented as follows:

S1.

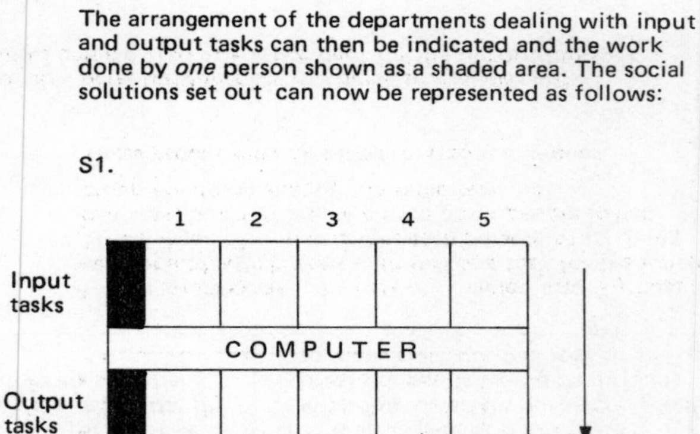

The office is divided into 5 large groups, which each deal with all stages of the work. The shaded area shows the work of *one* person in Group 1, who does all the different jobs necessary.

These groups may be called 'longtitudinal' groups.

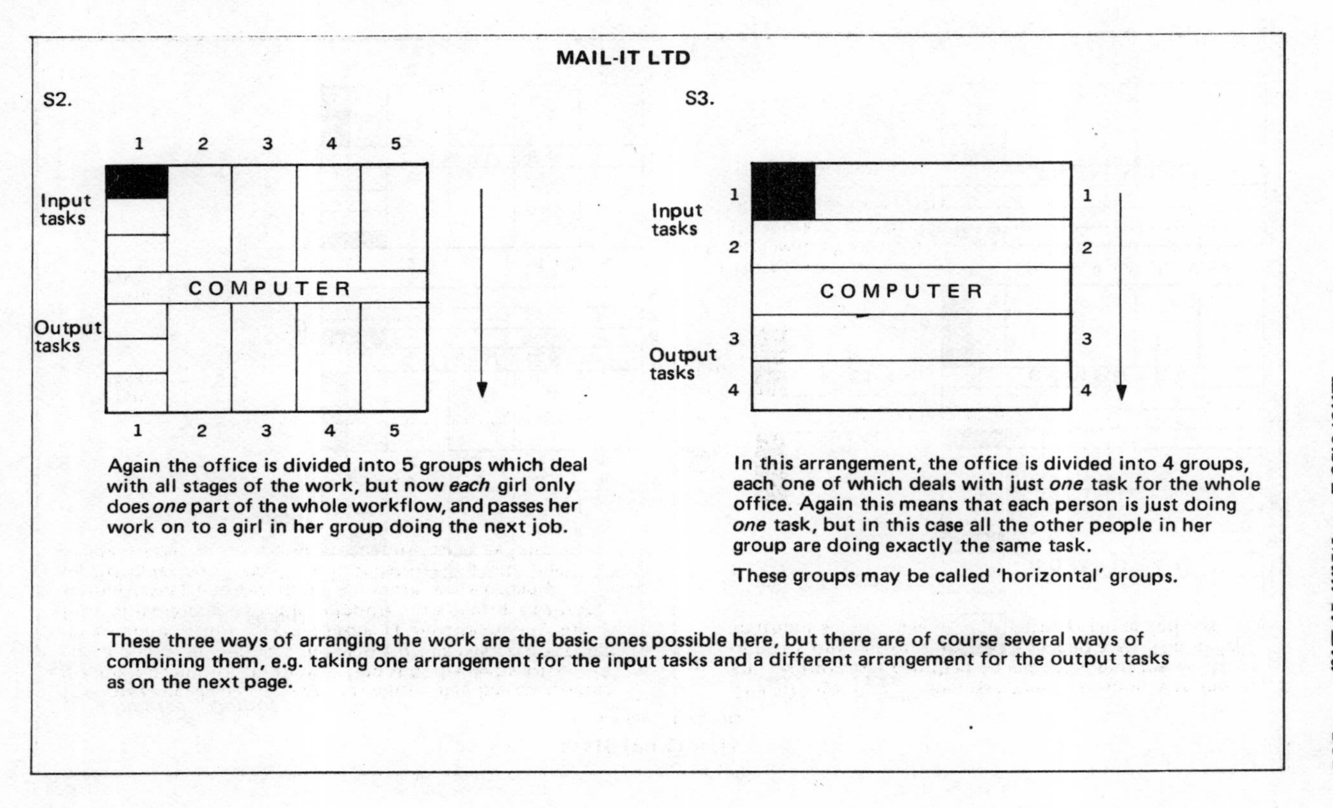

Again the office is divided into 5 groups which deal with all stages of the work, but now *each* girl only does *one* part of the whole workflow, and passes her work on to a girl in her group doing the next job.

In this arrangement, the office is divided into 4 groups, each one of which deals with just *one* task for the whole office. Again this means that each person is just doing *one* task, but in this case all the other people in her group are doing exactly the same task.

These groups may be called 'horizontal' groups.

These three ways of arranging the work are the basic ones possible here, but there are of course several ways of combining them, e.g. taking one arrangement for the input tasks and a different arrangement for the output tasks as on the next page.

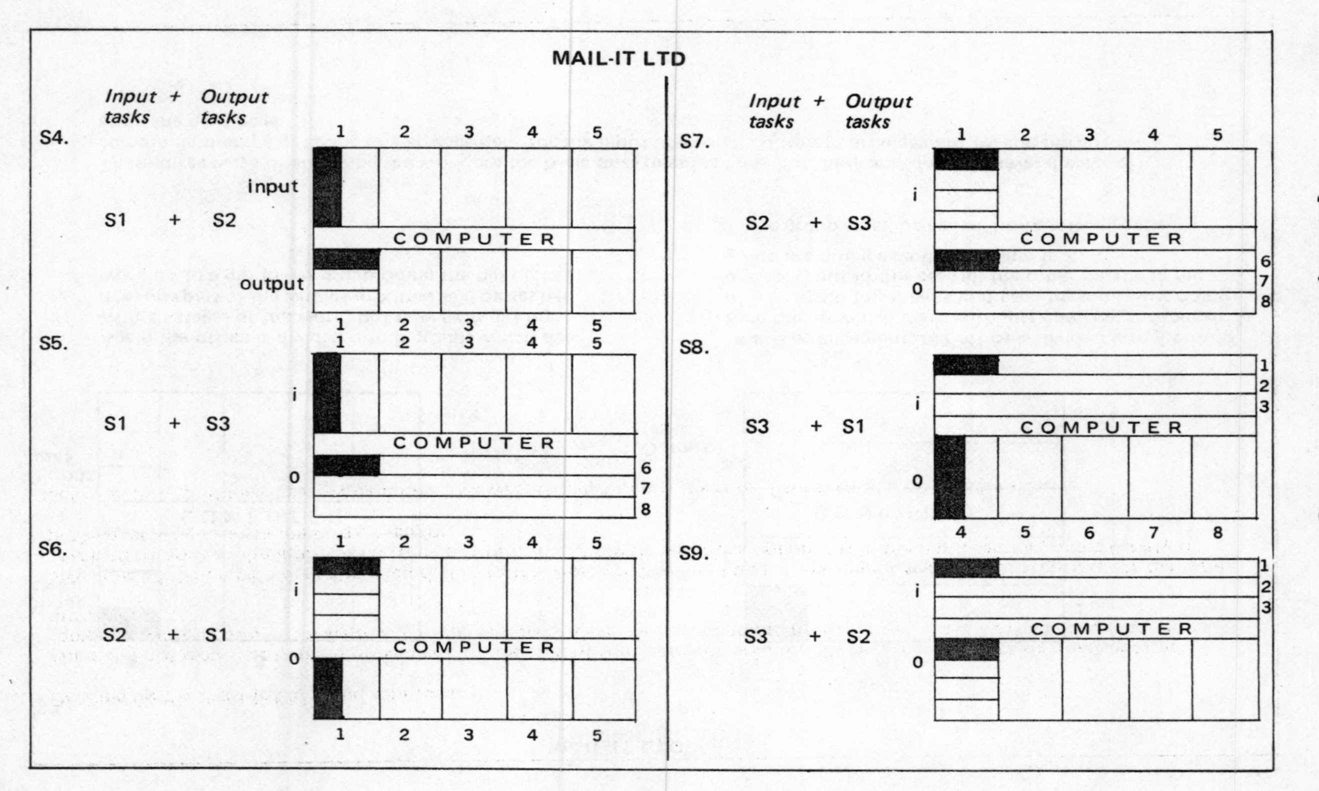
MAIL-IT LTD
S4.
Input + Output
tasks tasks
S1 + S2
input
output
COMPUTER
S5.
S1 + S3
S6.
S2 + S1
S7.
S2 + S3
S8.
S3 + S1
S9.
S3 + S2

MAIL-IT LTD

Drawing up the short-list of social solutions

There did not seem to be any improvement on the present situation with solution S3, and there were some disadvantages compared with the present system, e.g. in the opportunities for understanding the workflow, so S3 was excluded from the short-list.

Solutions S4, S5 and S7 were also excluded from the short-list, because these arrangements seemed inappropriate for the types of job to be done in the input and output sections; that is simple repetitive operations in the input section, and a variety of problem-solving tasks in the output section.

Solution S9 would have been a possibility, but was left out to keep the short-list down to a reasonable size.

MAIL-IT LTD

Technical solutions short-list		*Social solutions short-list*	
	Give brief description		*Give brief description*
T1.	Key to tape input. Batch processing update. Paper printout.	S1.	Office divided into groups with each girl doing all stages of work.
T3.	Paper tape input. Batch processing update. Paper printout.	S2.	Office divided into groups which deal with all stages of the work, but each girl does only one task.
T4.	Visual display units on-line for input. Batch processing update. Paper printout.	S6.	Input jobs done in longitudinal groups, but with each girl doing one job. Output jobs done in longitudinal groups — each girl doing all stages of the work.
T5.	Visual display units on-line for input. Batch processing update. Microfilm output for rejects.	S8.	Input done in horizontal groups — each girl does one job. Output done in longitudinal groups — each girl does all tasks.

MAIL-IT LTD

STEP 4 SET OUT POSSIBLE SOCIO-TECHNICAL SOLUTIONS

It is now necessary to merge your short-lists of technical and social solutions.

The essential thing is to see which technical and social solutions are compatible with one another and to eliminate any which cannot be fitted easily together.

Take each technical solution in turn and compare it with all the social solutions on your other short-list. Where the two solutions could be operated together, mark this combination down as a possible socio-technical solution on the next page. It may be that *all* your solutions are compatible with one another, in which case you can enter them all as possible socio-technical solutions.

Please use *your* social and technical solutions, not ours.

Do not rank the solutions in order of preference yet.

MAIL-IT LTD

Possible socio-technical solutions short-list

Description	*Ranking*	*Description*	*Ranking*	*Description*	*Ranking*
1.		4.		7.	
2.		5.		8.	
3.		6.		9.	

MAIL-IT LTD

STEP 5 RANKING SOCIO-TECHNICAL SOLUTIONS

The list of socio-technical solutions which *you* have just drawn up must now be ranked, before the most suitable system can be chosen. Turn back to Step 3, where you set out the efficiency and job-satisfaction advantages and disadvantages of the various solutions *you* put forward. Consider what you wrote on those pages and then try and rank the socio-technical solutions on the previous page from 1 to 9. If you find this rather difficult, you might like to award +1 to each advantage and –1 to each disadvantage until you achieve a score for each socio-technical solution, or to help differentiate further, you could award +2 to each major advantage and –2 to each major disadvantage.

When you think that you have achieved a satisfactory ranking, enter this against each socio-technical solution on the previous page. It should now be clear which socio-technical solution you consider the best, but it is necessary to check that this solution is *completely* satisfactory before accepting it as the most suitable system.

Our ranked socio-technical solutions are shown on the next page.

MAIL-IT LTD

Possible socio-technical solutions short-list

Description	*Ranking*	*Description*	*Ranking*	*Description*	*Ranking*
1. T1. Key to tape input. Batch-processing update. Paper printout. + S2. Office divided into groups which deal with all stages of work. Each girl does only one task.	9	4. T3. Paper tape input. Batch-processing update. Paper printout. + S2. Office divided into groups which deal with all stages of work. Each girl does only one task.	6	7. T4. Visual display units for input. Batch-processing update. Paper printout. + S2. Office divided into groups which deal with all stages of work. Each girl does only one task.	3
2. T1. Key to tape input. Batch-processing update. Paper printout. + S6. Input jobs in longitudinal groups. Each girl does one task. Output tasks done in longitudinal groups. Each girl does one task.	7	5. T3. Paper tape input. Batch-processing update. Paper printout. + S6. Input jobs in longitudinal groups. Each girl does one task. Output tasks done in longitudinal groups. Each girl does one task.	4	8. T4. Visual display units for input. Batch-processing update. Paper printout. + S6. Input jobs in longitudinal groups. Each girl does one task. Output tasks done in longitudinal groups. Each girl does one task.	1
3. T1. Key to tape input. Batch-processing update. Paper printout. + S8. Input done in horizontal groups. Each girl does one task. Output done in longitudinal groups. Each girl does all tasks.	8	6. T3. Paper tape input. Batch-processing update. Paper printout. + S8. Input done in horizontal groups. Each girl does one task. Output done in longitudinal groups. Each girl does all tasks.	5	9. T4. Visual display units for input. Batch-processing update. Paper printout. + S8. Input done in horizontal groups. Each girl does one task. Output done in longitudinal groups. Each girl does all tasks.	2

MAIL-IT LTD

CHECK YOUR CHOSEN SOCIO-TECHNICAL SOLUTION

Although you have already checked the technical and social solutions *separately,* it is now necessary to make sure that your combined socio-technical solution still meets the criteria laid down earlier. Turn back to Steps 1 and 2 and check that your chosen socio-technical solution meets the conditions set out there.

1. Does this solution meet *both* technical requirements and human needs?
2. Are sufficient resources available to achieve *both* the technical and social aspects of your chosen solution?
3. Do any of the constraints set out earlier make your chosen solutions impossible?
4. Does this solution meet both the technical and social objectives which were set out in Step 2?

If you are satisfied that your chosen socio-technical solution is still viable, proceed to Step 6 and prepare a detailed work design. If you find that your chosen solution has not met the criteria in Step 5, return to your short-list of socio-technical solutions and take the solution ranked next, and go through Step 5 again. Carry on until you find a socio-technical solution which meets the criteria in Step 5.

MAIL-IT LTD

STEP 6 PREPARE A DETAILED WORK DESIGN FROM YOUR CHOSEN SOCIO-TECHNICAL SOLUTION

Prepare a list and a description of all the tasks which clerks will be required to do if you implement your chosen socio-technical solution.

Now rank these tasks in order of simplicity. Is there a way of combining or arranging these tasks into jobs so as to give a balanced spread of required skills and complexity of tasks? If so, please describe your arrangement, using a simple flowchart.

Now check that this arrangement of tasks will create jobs which are as interesting and satisfying as possible.

1. Are there good feedback loops to each job, informing the employee on his performance?
2. Can the employee easily identify the targets he will have to achieve?
3. Are there clear boundaries between the different jobs so that the employee has a feeling of identity with his job?
4. Is the cycle time of the different tasks long enough to avoid a feeling of repetitive work but short enough to allow the employee to feel he is making progress with his work?

If you consider that the jobs you have created are as satisfying as they could be, while still achieving the technical objectives of the system then you may accept this socio-technical solution as your final solution.

If you have any doubts about the jobs which your chosen solution will create, go back to your short-list of socio-technical solutions, and take the solution which ranks next and proceed with Steps 5 and 6 again, until you are satisfied with your solution.

MAIL-IT LTD

Detailed work design

MAIL-IT LTD

Evaluation of solution

If you consider that the jobs you have created are as satisfying as they could be, while still achieving the technical objectives of the system then you may accept this socio-technical solution as your final solution.

If you have any doubts about the jobs which your chosen solution will create, go back to your short-list of socio-technical solutions, and take the solution which ranks next and proceed with Steps 5 and 6 again, until you are satisfied with your solution.

Set out below your arguments for selecting this particular solution:

MAIL-IT LTD

OUR SOLUTION IS SET OUT IN DETAIL ON THE FOLLOWING PAGES.

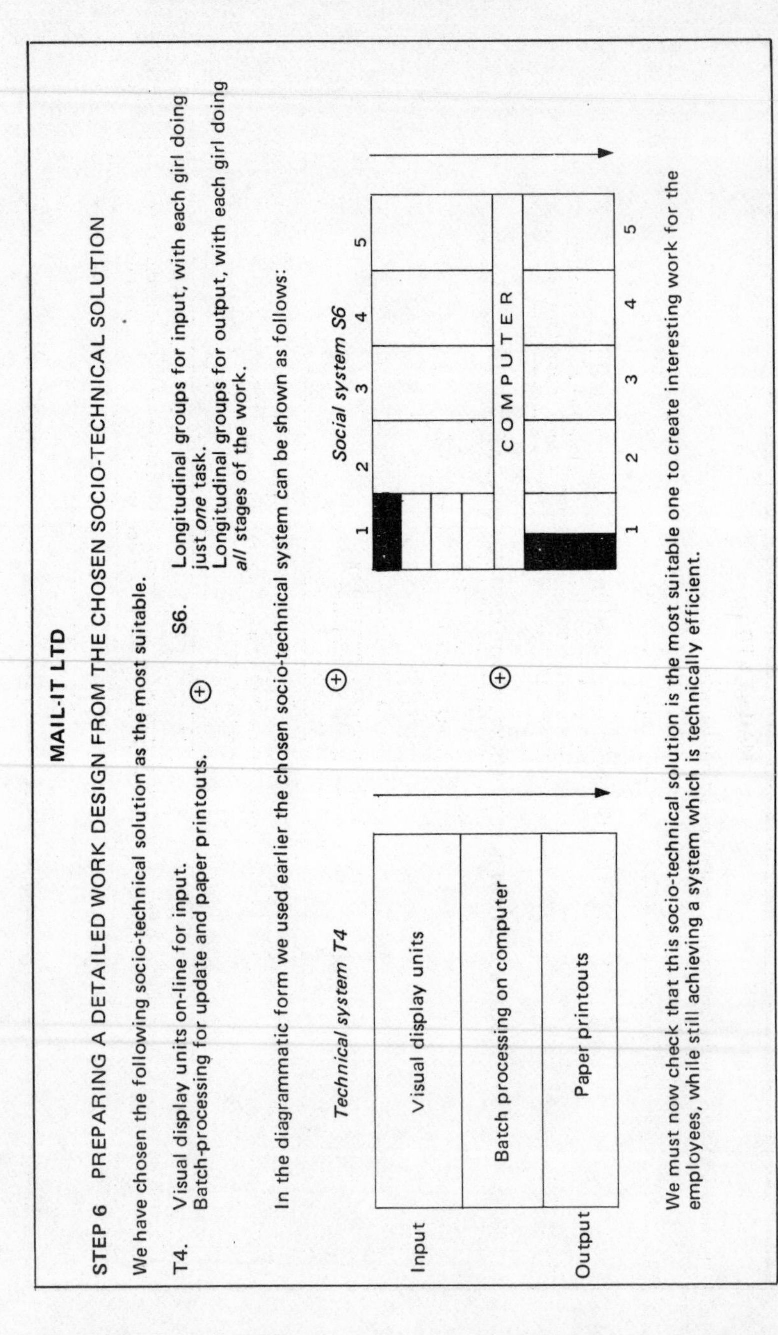

MAIL-IT LTD

STEP 6 PREPARING A DETAILED WORK DESIGN FROM THE CHOSEN SOCIO-TECHNICAL SOLUTION

We have chosen the following socio-technical solution as the most suitable.

T4. Visual display units on-line for input.
Batch-processing for update and paper printouts.

⊕

S6. Longitudinal groups for input, with each girl doing just *one* task.
Longitudinal groups for output, with each girl doing *all* stages of the work.

In the diagrammatic form we used earlier the chosen socio-technical system can be shown as follows:

We must now check that this socio-technical solution is the most suitable one to create interesting work for the employees, while still achieving a system which is technically efficient.

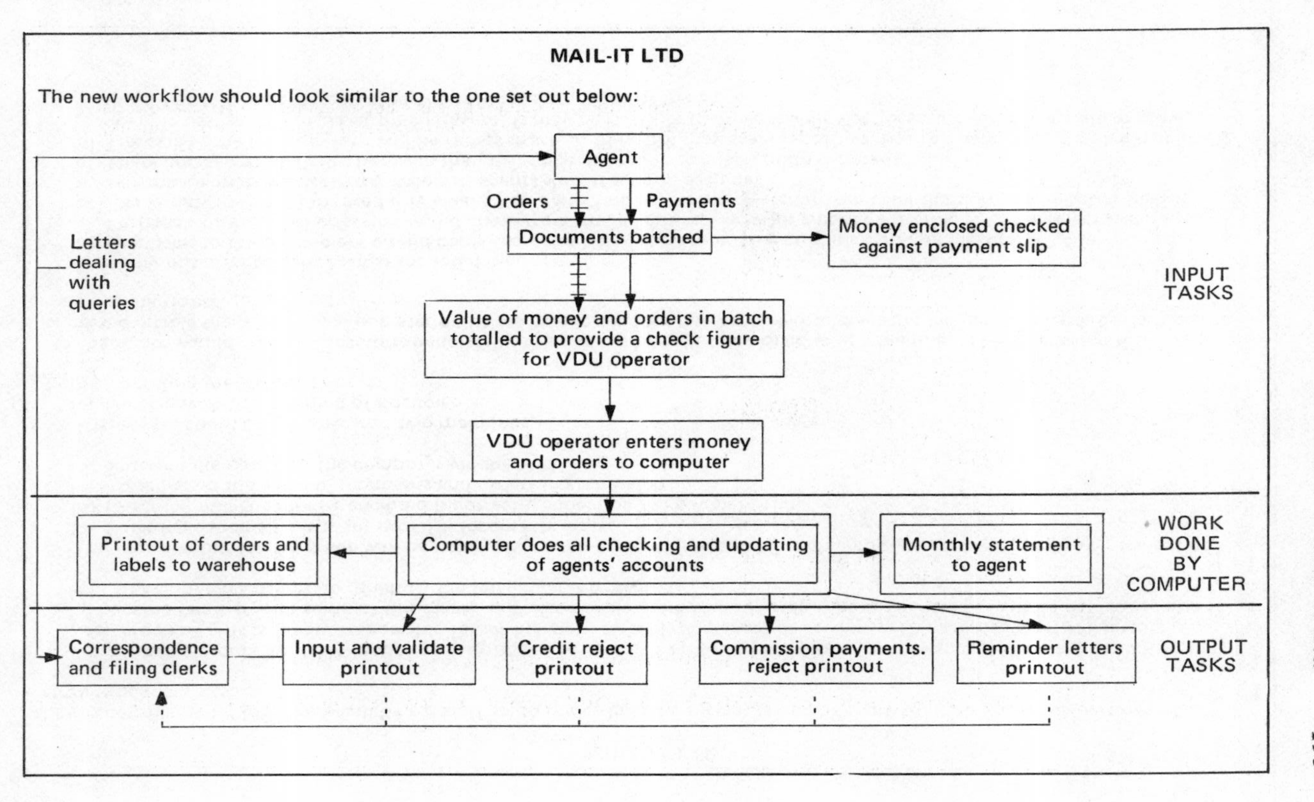
MAIL-IT LTD
The new workflow should look similar to the one set out below:
Agent
Orders
Payments
Documents batched
Money enclosed checked against payment slip
Letters dealing with queries
INPUT TASKS
Value of money and orders in batch totalled to provide a check figure for VDU operator
VDU operator enters money and orders to computer
Printout of orders and labels to warehouse
Computer does all checking and updating of agents' accounts
Monthly statement to agent
WORK DONE BY COMPUTER
Correspondence and filing clerks
Input and validate printout
Credit reject printout
Commission payments. reject printout
Reminder letters printout
OUTPUT TASKS

MAIL-IT LTD

Let us consider what jobs there will be for staff to do under the new system, and make sure that our chosen departmental arrangement is suitable.

1. Longitudinal groups for both input and output tasks will work very well, because we can divide all the accounts up, so that each group of girls is totally responsible for all the work on a particular set of accounts. This arrangement will enable staff to become familiar with some of the agents, and also enable them to understand the total workflow quite easily, because it will all be done by the group of which the individual is a member.

2. The input section of the group will be organised on the basis of one task for each employee and these tasks will consist of:
   (a) Batching documents into units of 10 documents each.
   (b) Checking that the money enclosed is correctly shown on the payment slip.
   (c) Totalling up the value of payments and orders in each batch.
   (d) Entering the data into the computer via the VDU.

   These tasks could be grouped into two main jobs,
   (a) Dealing with the batching of documents.
   (b) Operating the VDUs.

   These jobs would cover *all* accounts handled by the group of which the individual is a member. This arrangement of tasks would enable the VDUs to be operated full-time by experienced operators and so meet the technical constraint of efficient use of hardware.

   Checking this arrangement against the job design criteria:
   (a) Feedback to the batch clerk would come from the VDU operator who would pick up most errors.
   Feedback to the VDU operator would come from the output section who would be dealing with the reject printouts.
   (b) The targets for both jobs are fairly clear since clerks know how many accounts in total their group is responsible for, and approximately how many accounts agents should send in each day.
   (c) There are clear boundaries between the jobs, since the type of work is quite different.
   (d) The cycle time for both jobs will be rather short, so the work may seem repetitive, but this cannot be avoided if the best possible use is to be made of the hardware. It may be possible to move the girls between the jobs to give them some variety or to compensate in other ways for the boredom of the work.

MAIL-IT LTD

3. The output section of the group will have to deal with the four main printouts from the computer:
   (a) Input and validate — shows up any errors in the information entered via VDUs.
   (b) Credit reject — shows those agents who are not credit worthy for the goods they have ordered.
   (c) Commission reject — shows those commission claims from agents which contain errors or miscalculations.
   (d) Reminder letters — shows those agents who have fallen behind with payments.

   The chosen arrangement for the output section is groups in which the staff do all stages of the work. This arrangement will be well suited to the type of problem-solving work needed on the reject printouts. Each group will be responsible for the accounts of 20 000 agents, but within that, smaller subgroups could be set up looking after, e.g. 5000 agents each. The best way to organise the work might be for each sub-group to deal with the 4 printouts and the correspondence for its own agents. Each girl would deal with a printout right through — sorting out the reasons for rejection, taking a decision on what to do, and then either correcting the mistake, overriding the computer's decision, or writing to the agent explaining the reasons for the rejection. Within each group, the jobs could be graded so that new entrants started on the easiest job and gradually moved on to the other jobs. This would give the group flexibility during holidays and illness, and also provide a means of promotion and advancement, while retaining a stable work group. The first job would be input and validate, dealing with mistakes in the input data. The reminder letter is the next most straightforward job, followed by credit reject. The commission reject clerk's job is considered by Mail-It as important because it deals with commission which they regard as the agent's wages, and so likely to be a sensitive area to mistakes. Finally, the correspondence clerk has the most difficult job in the group, because the queries she receives could cover any aspect of the firm's work, and the public image of the firm depends to a large extent on how well customers think their queries are handled.

   Checking this arrangement against the job-design criteria:
   (a) Feedback to the output groups would come mainly from the agents, since they will write to complain if their queries are not dealt with properly, or if mistakes are made.
   (b) The targets for all the jobs are clear; they are to deal with all rejections, or letters for a particular group of agents. Clerks' own efforts can also help to reduce the number of letters they receive.
   (c) The boundaries between the jobs are clear, because each girl is dealing with a different printout from the computer — or with agents' letters.
   (d) The cycle time for the jobs will be just about right, since some queries will be dealt with quite quickly, while the more difficult ones will take longer, anything from 5 minutes to 30 minutes, and this should provide some variety for staff, and reduce feelings of repetition.

MAIL-IT LTD

STEP 7

The socio-technical solution which was chosen seems appropriate in that it will enable interesting jobs to be created, at the same time as permitting a technically efficient computer system to be set up. The use of visual display units provides a flexible input medium, and can also be used to provide reasonably satisfying jobs for those staff who like simple, machine-operating tasks. The arrangement of the output sections enables the rejections to be dealt with quickly, at the same time as creating interesting, problem-solving tasks for staff who prefer to undertake jobs which require some decision-taking and initiative. The group arrangement also provides a useful promotion and advancement ladder which could be linked to the payment system.

We can now finally accept this socio-technical solution as the best one possible for the work of the Mail-It Main Office.

6 The ETHICS method — Exercise 3

INTERNATIONAL BANK FOREIGN EXCHANGE DEPARTMENT

Background information

The Foreign Exchange Department of International Bank forms part of the Exchange and Money Market Division. This Division consists of the Chief Manager's establishment, the Dealing Room and the Exchange Department. The technical activities are performed in the Dealing Room; the Exchange Department is an administrative unit solely occupied with processing deals entered into by the Dealing Room.

The bulk of the transactions consist of buying and selling foreign currency by means of telephone communication to and from agents all over the world. Currency is also accepted on deposit from other banks and private customers and lent to banks and institutions. Dealing is organised by currency rather than transactions and the dealers are split up as follows:

- Dollar Section.
- Swiss franc Section
- Deutschmark Section.
- Continental Section for other currencies.
- Sterling Section.

The Dealing Room is staffed by *dealers* and *position clerks.* Position clerks keep a record of funds going in or out (the position) and know the value date of deals.

Exchange Department

It is this department with which we are primarily concerned in the exercise that follows, for it has many more routine activities than the Dealing Room. The work of the Department is essentially handling paper work but there is also an element of customer work and customer contact which people like. There is a strong belief that the talents of many clerks are being under-utilised by the present organisation of work.

Exchange Department staff consists of 68 clerks of whom 25 are part-timers. The grading structure is as follows:

- 3 Assistant Managers — Grade 7
- 2 Supervisors — Grade 6
- Section Leaders — Grade 5
- Deputy Section Leaders — Grade 4
- Checkers — Grade 3
- Coders — Grade 2
- Typists — Grade 2
- Filing Clerks — Grade 1

The main strengths of staff in Exchange Department are their ability to cope with heavy work loads under considerable pressure.

Their principal weakness is a lack of all-round banking knowledge. Due to economic circumstances in recent years, a number of relatively inexperienced staff have been promoted to jobs in which, traditionally, they would be required to have a higher level of banking than knowledge than they in fact possess.

This inexperience shows particularly at first-line management (Section Leader) level.

The present computer system

The processing is completely computerised with a batch computer system that has been in use since 1971. A copy of the deal ticket is used as an input document. This document is used to post all entries, the only manual operation being the movement of funds by telex, cable or mail transfer. The batch computer system was introduced to help the bank undertake high-volume routine processing in the foreign exchange field. Bank management was worried that in the future it would not be possible to attract enough staff to carry the work out manually with a rapidly expanding foreign exchange business. Computerisation provided an answer to this problem. However, the system has involved a great deal of routine coding work and has not proved an ideal application. One major problem has been the long delay in receiving information from the computer.

The principal objectives of the batch system were:

1. To speed up workflow.
2. To obtain more accuracy in arithmetical calculations.
3. To obtain better and more up-to-date management information.
4. To provide a quicker and more efficient service to customers.

The system aims to:

1. Record each transaction put through by dealers so that the bank's records and books can be maintained.
2. Provide basic documents for customers.
3. Generate diary notes to clerks on actions that have to be taken.
4. Provide management information that will assist bank decisions on how to proceed in the future.

It was also hoped that the system would eliminate a number of routine and boring clerical jobs. This did happen but these were replaced by even more routine and boring jobs associated with data processing by computer — for example the coding of input vouchers. Ledger work which provided many people with job satisfaction has virtually disappeared. There is also a great deal of routine checking work and a need to adhere to computer-input deadlines.

In human terms the batch system has had a detrimental effect. Clerical staff do not like it and the junior staff do not understand it. The general opinion amongst staff is that work has been made less interesting, although from the bank's point of view the introduction of the system has led to much greater work accuracy. The Departmental Manager's view is that the system produces too many lengthy reports and it is difficult for staff to take an interest in these.

This is not the Foreign Exchange Department's first computer system, as a system with similar functions to the present batch one was introduced in 1964. This was replaced because the machine was becoming obsolete and the advent of decimalisation meant that certain radical changes would have to be made to the program. It was therefore an opportune moment to change to another machine.

The proposed system review

The bank has now decided to review the use it is making of computers for foreign exchange purposes and to encourage the development of a new method of organising work in the Exchange Department, that provides staff with more job satisfaction.

There are a number of operational problems in the Department which need to be considered both in terms of the use of a computer and in terms of the reorganisation of work. These are:

Volume problems. Dollars are the dominant currency and these overwhelm the rest of the work.

Work peaking. Some dealers have two peaks in the day, 8.00 a.m.-10.00 a.m. and late afternoon. These affect the work of the Exchange Department.

Any new work system must of course provide for the tight controls that banking requires. The bank's rules are very explicit and do not allow for any discretion. This presents a design problem as it is not easy to ally a strict control system with staff opportunities for independence of mind and creative thinking.

At this stage there are no cost constraints which could prevent particular solutions being considered.

The manager of the Exchange Department hopes that an improved computerised system will enable him to reduce staff requirements by around 20, as well as enhancing the efficiency of the office by increasing the speed of throughput of work, and reducing the number of errors. At the same time he hopes to use this opportunity to improve the jobs which the clerks do, and he feels this will be beneficial both in terms of job satisfaction and in helping the bank to reduce turnover among the people they would like to keep, usually those clerks who prefer more demanding and interesting work.

Foreign Exchange are now about to begin the design of an improved work system which may incorporate a new computer system and they have decided to take technical, business and social objectives into account in the design process. Because they are interested in increasing employee participation in new bank developments, they have set up a systems design team which consists of a technical computer expert from management services, six senior clerks, five from the Exchange Department and one from the Dealing Room and the Manager of the Exchange Department.

Imagine that you are the systems design team of International Bank and follow through the ETHICS method. This should assist you to find the best possible socio-technical solution for Exchange Department.

A brief description of the work of a Foreign Exchange Department is set out below.

Foreign exchange dealing

Dealing staff are responsible to the management for buying and selling foreign currencies as and when required by all sections of the department and for the application of rates of exchange to all transactions involving currency operations. The dealers are responsible for the department's currency 'positions' and for the maintenance and distribution of the bank's currency balances with its overseas correspondents.

In any free market, supply and demand are the principal factors in determining price. These factors also play a major part in arriving at the current exchange value of any currency.

If there is a demand for US dollars against sterling then — other things being equal — the tendency will be for the dollar rate to 'harden'. In exchange for £1 a buyer will get fewer US dollars. In consequence the rate for the US dollar which was originally quoted as 2.4305-2.4315 might appreciate to 2.4275-2.4285.

Exchange rates can, and do, respond very quickly to the effect of noticeable disequilibrium of supply and demand and also to news of political or economic importance, whether at home or abroad.

Although bankers refer to the London Foreign Exchange Market, there is no central market as such. The 'Market' consists of dealers, operating in the foreign departments of the banks, and brokers whose specialist function is to bring together buyers and sellers among the dealers.

Brokers do not carry stocks of currency and make no deals with the public — unlike jobbers in the Stock Exchange. For bringing together a buyer and a seller a broker is remunerated by receiving brokerage or commission, charged at predetermined rates.

Dealers buy or sell in the market only through brokers and transactions are done solely by means of telephone. There are private lines linking banks to brokers — thus one foreign department may have lines to five or six brokers.

A deal

A deal may be concluded in the following manner:

1. The dealer of International Bank enquires of Broker A the latest quotation for US dollars and is told that this is 2.4305-2.4315
2. International Bank says that it has $500000 for sale.
3. Broker A telephones Bahamas Bank, who he knows are wanting dollars and will offer them at 2.4305.
4. Bahamas Bank accepts the rate.
5. The broker advises Bahamas Bank that the dollars will be provided by International Bank.

6. The two banks exchange instructions.
7. The deal is ratified by an exchange of contract notes between the two parties buying and selling. These indicate the name of the broker.
8. Two business days later Bahamas Bank pays International Bank the sterling amount involved and International Bank supplies the dollars.

There are many important foreign exchange markets in the principal cities overseas such as Paris, Zurich, Frankfurt, New York and Tokyo. Dealers in London can apply to these for rates and make a profit if the rate for a currency gets temporarily out of alignment between the two centres. This has the effect of speedily correcting inequalities and rates tend to be very close, whether quoted in the London Market or in any of the cities just mentioned.

There is a constant movement of rates due to the supply and demand of currencies, the variation of interest rates between the two centres, and other pressures such as threats of war, industrial action, or sanctions where, for example, oil producers arbitrarily limit supplies to their international customers. Dealers therefore look for their profits from the 'spread' or margin between their buying and selling rates for each currency.

In addition to profit from 'spread', dealers charge customers Exchange Commission at a rate of £1 per mille. with a maximum on any one transaction of £10.

The dealer's position

Each deal of any size with the public is covered, as it is undertaken, either by a compensating deal on the market or, if the dealer is fortunate, by a simultaneous offsetting transaction with the public — i.e. a 'marry'.

At the commencement of each day's trading the dealer has a *position sheet* for each currency, showing whether he will commence trading in an over-bought or an over-sold state — or 'long' or 'short'. Each deal concluded is recorded on the position sheet and the dealer therefore has before him at all times an up-to-date record of his position in every currency.

A London dealer will never allow himself to have a large uncovered position in any currency as, with the constant movement of rates, he could suffer loss, or reduced profit, if the rates moved adversely before he took corrective steps to balance his books. (The Bank of England gives to each UK bank limits in respect of uncovered positions which they may not exceed. Returns are called for weekly to ensure that the limits imposed are not violated.)

In maintaining their books dealers have to take great care that they are covered in a currency, not only as regards a currency but also by time. For instance, if a dealer buys from several customers cheques drawn on American banks and totalling $100000, he may sell a similar amount by telegraphic transfer either to another customer or through the market and so balance his position. The dollars telegraphed will be debited to the account in New York two days later whereas the cheques sent to New York by mail or, as is usual now by courier, could conceivably take seventy-two hours or longer before the same amount is credited, thus possibly creating an overdraft on which interest could be charged. The dealer must therefore continually check that he has a free working balance available to compensate for differing value dates in his operation.

Running a position

In some overseas centres dealers deliberately remain, for several days, with over-bought or over-sold positions in certain currencies in the hope that exchange rates will move in their favour and permit them to make additional dealing profit. This practice of 'running a position' is not encouraged in this country as it is pure speculation and UK dealers are strictly controlled as to their limits by Bank of England directions and authorities.

The uncovered balance, or open position, at the close of business as shown in the dealer's position sheets should, after certain adjustments, agree with the Nostro balances in the ledger section (see page).

It is continually necessary for the dealer to ensure that

sufficient balances are maintained with correspondents and to instruct the remittances sections as to which correspondent is to be utilised in carrying out the customer's instructions.

Forward dealing

Where a transaction is carried out in foreign currency without the assistance of a currency account, the only effective way in which the exchange risk can be eliminated is by forward exchange. Here the trader, by means of an exchange contract with his banker, fixes at once the rate at which a future exchange deal will be effected. He then knows definitely the amount of sterling he will receive or have to give in connection with the transaction. Forward exchange may be cheaper or dearer than spot, depending on market conditions. It may be mentioned here that in certain circumstances a trader may purchase funds required for a future transaction spot and have them held for his account by an authorised bank. Whether it will pay him to do so will depend on whether his financial position is sufficiently liquid to allow him to lie out of his sterling for the period in question, and the cost of the cover desired.

A dealer is prepared to buy or sell forward exchange either for a fixed date, or for delivery during a stated period at the customer's option. There is no option on the part of the customer as to whether he will complete the transaction or not, as in a Stock Exchange option, but the option is merely as to when the transaction is to be completed within a given period. Thus 'option June and July' means that delivery can take place any time between 1 June and 31 July, but the transaction must be completed by the latter date.

When a dealer undertakes a forward deal, he must cover his deal just as he would were it spot, although the method he uses will vary with the type of transaction and the state of his position in the currency involved. In covering, the dealer undertakes a definite obligation on behalf of the bank, which must be fulfilled whether the deal with the customer is completed or not. It is essential, therefore, that forward deals with customers be put on a formal basis and that an agreement under hand, in which the terms of the deal are set out in full, and the customer's obligations clearly stated and accepted, should be taken from the customer. Nor should forward deals be undertaken for any but the most reliable customers. In large deals, a sterling deposit of a suitable margin may be required unless the standing of the customer is quite undoubted.

Accounting

Ledgers used in most foreign exchange departments conform to a general pattern. In most British banks these are now fully computerised.

The main books of accounts are normally three in number:

1. The *Nostro Ledger* in which are recorded the details of the bank's currency accounts with banks overseas.
2. The *Vostro Ledger* in which are recorded the sterling accounts held in the name of the bank's foreign correspondents.
3. The *Customers' Currency Accounts Ledger* in which is recorded foreign currency balances held on account of branch customers.

Nostro Ledger

Currency accounts are maintained with correspondents in all the principal financial centres throughout the world. These accounts are fed with currency items which the Foreign Exchange Department buys from or collects on behalf of the bank's customers or with funds purchased in the Foreign Exchange Market. Against these accumulated balances the Foreign Exchange Department is able, for example, to effect cable or mail transfers or issue drafts. It is the Dealing Section's responsibility to ensure that adequate balances are maintained in any given account to meet debits by the correspondent bank; otherwise the Department could be required to meet overdraft interest.

Dealers cannot, with any degree of certainty, be aware what items have been credited or debited to the account abroad. For example, the correspondent may have issued a draft of US $10000 and on the date of issue have credited the Department's account. The dealer will be unaware of this increase in his dollar balance until the foreign bank's draft advice reaches the Department, or, if no advice is issued, until the draft is presented by the holder for payment. An adequate working balance must always be kept.

The Department is a customer of its overseas correspondents and the accounts which it has in Nostro Ledger are, in fact, the equivalent of bank statements maintained to confirm to the Department the state of its accounts overseas. If it maintains an account with a New York bank, then that bank is the Department's debtor to the extent of the balance. Entries appearing in the Department's Nostro Ledger are the reversal of those appearing in its correspondent books. Items which increase Nostro Ledger balance are *debits* while entries which reduce the balance are *credits*.

The Nostro Ledger is posted from entry slips prepared in both currency and sterling, the latter being calculated at the rate of exchange provided by the dealing section. If currency is being fed into and drawn from the Nostro Accounts at continually varying rates of exchange, it will be evident that the market value of the currency balance and the sterling balance shown in the ledger are unlikely to agree. The sterling difference represents the dealing profit or loss over any given period.

Long or short on net positions. Periodically the Department's net position in each currency is worked out and will confirm whether the dealer is overbought ('long') or oversold ('short'). If, at the end of the period, it was oversold in US dollars the Department would, in theory, require to purchase this amount at the Foreign Exchange Market selling rate of the day. The sterling amount required to make this purchase is compared with the book value shown in the ledger which, if greater, would produce a quantified profit in US dollar-dealing which would then be carried to a Profit and Loss Account to the debit of the US Nostro Accounts. After the profit or loss for each currency has been extracted, the sterling balance remaining will be the equivalent of the currency balance at the current market rate.

Account reconciliation. A correspondent bank may issue a draft on the first day of the month and at the same time credit the Department's currency account. The corresponding debit in the Department's Nostro Ledger might not be made until three weeks later when the draft is presented for payment. Conversely the Department could debit its Nostro Account on the first day of the month and air mail on that date to New York, dollar cheques totalling $150000. It might be the 5th of the month before a credit was passed over to the account in New York. Such differences in entry dates give rise to continued and complicated reconciliation procedures which are of basic importance to the efficient running of a Foreign Exchange Department. The task of reconciling is made somewhat easier by reason of the fact that a bank abroad will normally forward an advice (or value date) for each debit or credit it passes. The ledger keeper in the Foreign Department will note the date these *value dates* are issued against the date in his Nostro Ledger.

This will assist in two ways:

1. Marking off items in a statement issued at a later date will be simplified.
2. If no 'value date' is entered against an entry it will draw attention to the fact that a particular transaction may not, for a variety of reasons, have been finally effected and will place the ledger clerk on notice to make precautionary enquiries.

Vostro Ledger

This corresponds to the Current Account Ledger in a branch bank except that the individual accounts in it are in the name of the bank's overseas correspondents and will almost exclusively be designated as External accounts and subject to the terms of the UK Exchange Control. It is quite normal for individual credit and debit entries over such accounts to be advised to the correspondent bank twice daily by advice note followed at short intervals by computerised statements. Every Vostro Account maintained in a Foreign Exchange Department has a corresponding Nostro Account in the books of an overseas correspondent.

Customers' Currency Accounts Ledger

A non-resident of the UK or a resident individual or company holding Exchange Control authority may deposit amounts of foreign currency with a Foreign Exchange Department. These balances are recorded in the Customers' Accounts Ledger and are sometimes referred to as Loro accounts ('their' or 'third party' accounts). Such funds are held by the Department in its accounts with overseas correspondents but 'earmarked' for the exclusive use of the customer.

The Department, in agreement with a customer, may place currency balances held on deposit at call or for a specified period in order that interest may be obtained.

Forward sales and forward purchases

The Department will also keep records which show at any time:

1. The Department's forward position in each currency.
2. The Department's total liability under forward contracts.
3. The liabilities of each customer under forward contracts.
4. The maturity dates of individual forward contracts.

Definitions

Fixed deposits. A fixed deposit is an amount of money lent by or to the bank for a fixed period of time at a given rate.

Processing involves making the necessary entries on customer accounts. For example:

On starting date	Debit agent.	Credit fixed deposit account.
On maturity date	Debit fixed deposit account with principal amount.	Credit agent with principal amount.
	Debit currency interest (payment is made with interest).	Credit agent with interest.
Fixed loans.	Similar to fixed deposits.	

Market deposits are deals between banks either in London or abroad and form part of the London Money Market activity. The only restriction on dealing is the limit established for each bank by International Bank.

Customer deposits are deposits from private entities, either persons or companies. Under the existing exchange regulations issued by the Bank of England, residents of this country are not allowed to hold foreign currencies, except where they have received special permission from the Bank of England.

Agents are foreign banks which act as agents for International Bank.

Deposit guarantees are instructions from an overseas branch or a customer. If an overseas branch is giving a customer a loan — e.g. in Argentine currency — he may borrow this against dollars which International Bank holds to his credit. International Bank acts as guarantor.

INTERNATIONAL BANK

STEP 1 IN THIS EXERCISE WE ARE CHANGING OUR DIAGNOSTIC PROCEDURES SOMEWHAT – AS YOU WILL SEE WHEN YOU READ THE SECTION ON HUMAN NEEDS ON THE NEXT PAGE.

INTERNATIONAL BANK

STEP 1 DIAGNOSIS

Specify technical requirements and human needs.

Technical requirements

Physical factors To produce an improved, computerised system of work which reduces staff numbers required, handles the processing of deals efficiently, reduces the possibility of error.

Rate factors The system must provide an efficient service to customers by producing advice notes quickly.

Control factors There must be checking mechanisms to ensure that errors are quickly detected.

Human needs

You are now familiar with our diagnostic questionnaire and with the analysis of human needs which is derived from the questionnaire data. In this exercise we start by giving you the analysis of human needs which is set out on the next 3 pages. Please read it through carefully.

(Asterisk or make written comments as appropriate)

INTERNATIONAL BANK
Analysis of Human Needs

	Satisfactory aspects	*Should be incorporated into new system?*	*Unsatisfactory aspects*	*Must be improved in new system*	*Suggestions on how this can be achieved*
Knowledge 'fit'					
Use of knowledge			Skills and knowledge under-utilised.	*	Creation of larger jobs requiring greater use of different skills.
			Work not interesting.	*	
			Work not difficult enough.	*	
Self-development			Not enough opportunities for self-development.	*	Job progression from simple to increasingly difficult jobs.
Psychological 'fit'					
Status	Has status.	Yes.			
Responsibility	Has responsibility.	Yes.			Try and provide same amount of responsibility.
Recognition			Management not seen as recognising good work.		Through management-education programme associated with new system.
Job security	Has job security.				
Social relationships	Good social relationships.	Yes.			
Promotion	Good promotion opportunities.	Yes.		*	Promotion to other departments.
Achievement	Clerks feel achievement if they do their work well.	Yes.		*	This should look after itself if more interesting jobs are provided.

INTERNATIONAL BANK
Analysis of Human Needs

	Satisfactory aspects	*Should be incorporated into new system?*	*Unsatisfactory aspects*	*Must be improved in new system*	*Suggestions on how this can be achieved*
Efficiency 'fit'					
Salary	Level of pay.	Yes.			
Amount of work			Clerks feel overloaded.	*	Ensure workload is not increased. Feelings of overload may be due to low interest.
Standards	Standards seen as appropriate.	Yes.			Note existing level of accuracy is approved of by staff, but not by management.
Controls	Satisfactory, but errors still occur.	Yes.		*	See new system ensures fewer missed errors.
Supervision	Control of work by supervision seen as satisfactory.	Yes.			Ensure supervisory control stays at same level as now or improves.
Information and materials	Reasonably satisfactory.	Yes.			Ensure information relevant to task stays at same level as now or improves.
Task-structure 'fit'					
Work variety			Not enough variety in the work.	*	Larger jobs involving the use of more skills.
Initiative			Not enough scope for initiative.	*	Jobs which incorporate some uncertainty so that choices have to be made.
Judgement and decisions	No strong feelings.		No strong feelings.		
Pressure			Too much pressure	*	Better distribution of work loads throughout day will reduce pressure.
Targets	Satisfaction with targets.	Yes.			Ensure that targets and feedback are incorporated into new system.

INTERNATIONAL BANK
Analysis of Human Needs

	Satisfactory aspects	*Should be incorporated into new system?*	*Unsatisfactory aspects*	*Must be improved in new system*	*Suggestions on how this can be achieved*
Task-structure 'fit' (continued)					
Dependency on others	Clerks are dependent on the accuracy of others but there are no strong feelings about this.		See satisfactory aspects.		
Autonomy			More autonomy is required.		Larger jobs with well defined beginnings and ends and few constraints on how work is done.
Task identity	Job seen as important.	Yes.	More important jobs wanted.	*	Again, through larger, more meaningful sets of tasks.
Ethical 'fit'					
Company ethos	The bank is perceived as a benevolent employer that values experience in its staff.	Yes.			Ensure that nothing happens to alter clerks' attitudes to the bank.
Communication	Good at department level.	Yes.			Ensure that communication within the bank continues to be good.
Consultation			Better consultation required on major changes such as the new computer system.		Establish improved consultation mechanisms to cope with change.
Involvement	There is a low level of involvement with the bank.				

INTERNATIONAL BANK

Identifying different interests

On the basis of this analysis of human needs a number of human objectives have been set by the systems design team of which you are a member. Because the Manager of the Exchange Department is also a member of this team, he has insisted on a number of business objectives being set which the new work system must be designed to achieve. The technical computer expert has also included a number of technical objectives related to the use of the computer.

PLACE YOURSELF, IN TURN, IN THE ROLES OF THE COMPUTER EXPERT, THE USER MANAGER (MANAGER OF THE EXCHANGE DEPARTMENT) AND THE CLERKS OF THE USER DEPARTMENT (EXCHANGE DEPARTMENT) AND RANK EACH OBJECTIVE ACCORDING TO THE IMPORTANCE WHICH YOU THINK EACH WOULD ATTACH TO IT. THE FIRST TWO OBJECTIVES ARE WEIGHTED FOR YOU.

INTERNATIONAL BANK

Weight technical, human and business needs and objectives

Please rank the objectives listed below on a scale from 1—5 (5 = very important, 1 = of little importance) according to the importance each occupational group would attach to it.

	Name of need or objective	*Computer Expert*	*User Manager*	*Clerks from user dept*
1.	Provide incentives for high quality work.	3	5	3
2.	Ensure jobs are designed to provide variety, targets, judgement, task identity.	5	5	5
3.	Ensure control system is designed to provide incentives, feedback on performance, control with individual or group so that there is responsibility for own effort.			
4.	Provide effective training for change.			
5.	Ensure organisation of department and promotion system permits the continual development of skills and knowledge.			
6.	Ensure staff have materials and information necessary to carry out job efficiently.			
7.	Provide a pleasant and efficient working environment.			
8.	Increase scope for use of initiative and own ideas.			
9.	Maintain job security.			
10.	Maintain or increase earning levels.			
11.	Improve staff feeling of identity with the bank.			
12.	Improve information to staff from management.			
13.	Ensure high level of co-operation between users and Management Services.			
14.	Ensure advice notes sent to customer same day as deal is put through.			
15.	Reduce number of errors.			
16.	Improve control of dealing in overseas branches.			
17.	Ensure swift, helpful replies to customer queries.			
18.	Improve market penetration.			
19.	Improve productivity at current volumes.			
20.	Avoid developments with unsatisfactory financial return.			

INTERNATIONAL BANK

Weight of technical, human and business needs and objectives (continued)

	Name of need or objective	*Computer Expert*	*User Manager*	*Clerks from User dept.*
21.	Adapt more easily to changing conditions.			
22.	Speed up management information.			
23.	Hold more comprehensive data bases.			
24.	Simplify procedures for agents, customers and clerks.			
25.	Improve promotion opportunities.			
26.	Decrease work pressure.			

Priority needs and objectives

Please write below the objectives to which you have given 5 points. You should pay particular attention to ensuring that these are catered for in the design of the computer system. Other needs and objectives must not be ignored.

Computer Expert	*User Manager*	*Clerks*

INTERNATIONAL BANK

STEP 2 DESCRIBES THE CONSTRAINTS WITHIN WHICH YOU MUST WORK WHEN DESIGNING THE NEW SYSTEM. IT ALSO TELLS YOU THE RESOURCES THAT ARE AVAILABLE TO ASSIST YOU.

INTERNATIONAL BANK

STEP 2 SOCIO-TECHNICAL SYSTEMS DESIGN

Identify constraints

Identify technical and business constraints on the design of the system

The system must be sufficiently robust to reduce the amount of breakdown time to the minimum. There should also be adequate backup and recovery procedures, to maintain a high level of service to the customer.

The hardware must be used as economically as possible, that is, the maximum possible use must be made of the hardware which is compatible with the social constraints and social objectives.

Customer service must always be given priority in decisions taken.

Identify social constraints on the design of the system

The overall number of staff must be reduced (by 15 to 20 if possible), but there must be no redundancy.

The jobs created must be appropriate for the skill levels of the existing staff, bearing in mind that they feel their skills are underutilised at present.

INTERNATIONAL BANK

Identifies resources

Identify resources available for the technical system

The existing Management Services Department has experienced systems analysts and programmers available, though more will need to be recruited.

Assistance may also be available from the computer manufacturer supplying the hardware.

The bank has allocated sufficient funds for a major system improvement and is prepared to leave the hardware decision to the experts, so long as the budget is not exceeded.

Identify resources available for the social system

Personnel and training facilities are available for any retraining that is required.

Assistance is also available from the Organisation and Methods Department.

INTERNATIONAL BANK

AGREEING ON COMMON OBJECTIVES

ON THE BASIS OF THE WEIGHTS WHICH YOU IN YOUR DIFFERENT ROLES HAVE ALLOTTED TO THE VARIOUS OBJECTIVES, YOU MUST NEXT IDENTIFY A NUMBER OF TECHNICAL AND BUSINESS OBJECTIVES WHICH YOU BELIEVE WILL BE ACCEPTABLE TO THE WHOLE SYSTEMS DESIGN TEAM. I.E. NO GROUP HAS GIVEN THEM A WEIGHT OF LESS THAN THREE.

WRITE THESE OBJECTIVES IN ON THE NEXT PAGE.

INTERNATIONAL BANK

Specify priority technical and business objectives

(1) and (2) are obligatory objectives.

(NB Only objectives which can be achieved through the way the computer system is designed should be listed.)

To introduce a computer system to deal with the work of the Foreign Exchange Department. This system should:

1. Reduce staff requirements which are becoming increasingly difficult to meet because of a shrinkage in the labour market.
2. Improve the level of service to customers.
3.
4.
5.
6.
7.
8.
9.
10.

INTERNATIONAL BANK

AGREEING ON COMMON OBJECTIVES

NOW IDENTIFY A NUMBER OF HUMAN OBJECTIVES WHICH ALL THREE GROUPS IN THE SYSTEMS DESIGN TEAM AGREE ARE IMPORTANT. WRITE THESE IN ON THE NEXT PAGE.

INTERNATIONAL BANK

Specify priority human objectives

(NB Only objectives which can be achieved through the way the computer system is designed should be listed.)

1.

2.

3.

4.

5.

6.

7.

CHECK THAT YOUR TECHNICAL AND HUMAN OBJECTIVES ARE COMPATIBLE.

INTERNATIONAL BANK

OUR WEIGHTS AND OBJECTIVES ARE SHOWN ON THE FOLLOWING PAGES.

INTERNATIONAL BANK

Weight of technical, human and business needs and objectives

	Name of need or objective	*Computer Expert*	*User Manager*	*Clerks from user dept.*
1.	Provide incentives for high-quality work.	3	5	3
2.	Ensure jobs are designed to provide variety, targets, judgement, task identity.	5	5	5
3.	Ensure control system is designed to provide incentives, feedback on performance, control with individual or group so that there is responsibility for own effort.	5	5	5
4.	Provide effective training for change.	3	5	5
5.	Ensure organisation of department and promotion system permits the continual development of skills and knowledge.	3	5	5
6.	Ensure staff have materials and information necessary to carry out job efficiently.	2	3	5
7.	Provide a pleasant and efficient working environment.	3	5	5
8.	Increase scope for use of initiative and own ideas.	1	5	5
9.	Maintain job security.	1	2	5
10.	Maintain or increase earning levels.	1	1	5
11.	Improve staff feeling of identity with the bank.	1	5	3
12.	Improve information to staff from management.	3	5	5
13.	Ensure high level of co-operation between users and Management Services.	5	3	3
14.	Ensure advice notes sent to customer same day as deal is put through.	3	5	3
15.	Reduce number of errors.	2	5	2
16.	Improve control of dealing in overseas branches.	2	5	2
17.	Ensure swift, helpful replies to customer queries.	2	5	3
18.	Improve market penetration.	3	5	3
19.	Improve productivity at current volumes.	3	5	2
20.	Avoid developments with unsatisfactory financial return.	5	5	1

INTERNATIONAL BANK

Weighting of technical, human and business needs and objectives (continued)

Name of need or objective	*Computer Expert*	*User Manager*	*Clerks from user dept.*
21. Adapt more easily to changing conditions.	5	5	5
22. Speed up management information.	5	5	1
23. Hold more comprehensive data bases.	5	3	1
24. Simplify procedures for agents, customers and clerks.	3	5	3
25. Improve promotion opportunities.	1	3	5
26. Decrease work pressure.	3	3	5

Priority needs and objectives

Please write below the objectives to which you have given 5 points. You should pay particular attention to ensuring that these are catered for in the design of the computer system. Other needs and objectives must not be ignored.

Computer expert	*User manager*	*Clerks*
2. Design jobs to provide variety, etc.	2	2
3. Control systems to provide incentives, feedback, group control.	3	3
	4. Training for change.	4
	5. Development of skills and knowledge.	5
		6. Information to work efficiently.
	7. Pleasant work environment.	7
	8. Better opportunities for use of initiative.	8
		9. Maintain job security.
		10. Maintain/increase earnings.
	11. Improve staff identity with bank.	
	12. Improve information to staff.	12
13. Ensure co-operation of other groups with management services.	Also 14, 15, 16, 17, 18, 19 (efficiency objectives).	
20. Ensure financial return.	20	
21. Better adaptation to change.	21	21
22. Faster management information.	22	25. Improve promotion.
23. Larger data base.		26. Decrease work pressure.

INTERNATIONAL BANK

Priority technical and business objectives

To introduce a computer system to deal with the work of the Foreign Exchange Department. This system should:

1. Reduce staff requirements which are becoming increasingly difficult to meet because of a shrinkage in the labour market.
2. Improve the level of service to customers.
3. Ensure fast, accurate despatch of advice notes the same day the deal is made.
4. Make control of the Foreign Exchange Department easier and less onerous.
5. Simplify procedures for agents, customers and clerks.
6. Improve market penetration.
7. Assist the Exchange Department in becoming more flexible and adaptable.

INTERNATIONAL BANK

Priority human objectives

1. Jobs to be designed so as to provide variety, targets, opportunity to use judgement and initiative, task identity.
2. Control system to be designed to provide performance incentives, feedback, responsibility through individual and group control of own work effort and quality.
3. Organisation of department should assist progressive development of clerks' skills and knowledge.
4. Provide pleasant, efficient work environment free from excessive pressure.
5. Improve information to staff.
6. Maintain or increase earnings.

CHECK THAT YOUR TECHNICAL AND HUMAN OBJECTIVES ARE COMPATIBLE

INTERNATIONAL BANK

PRE-CHANGE WORK STRUCTURE

The next 12 pages contain an organisation chart of the Exchange Department, the work flow of those sections which contain the most clerks and handle the work that is most central to the operation of the Exchange Department, and short descriptions of typical jobs. Use these to gain an understanding of the pre-change work structure but remember your task is to create a *new* structure that increases job satisfaction and efficiency.

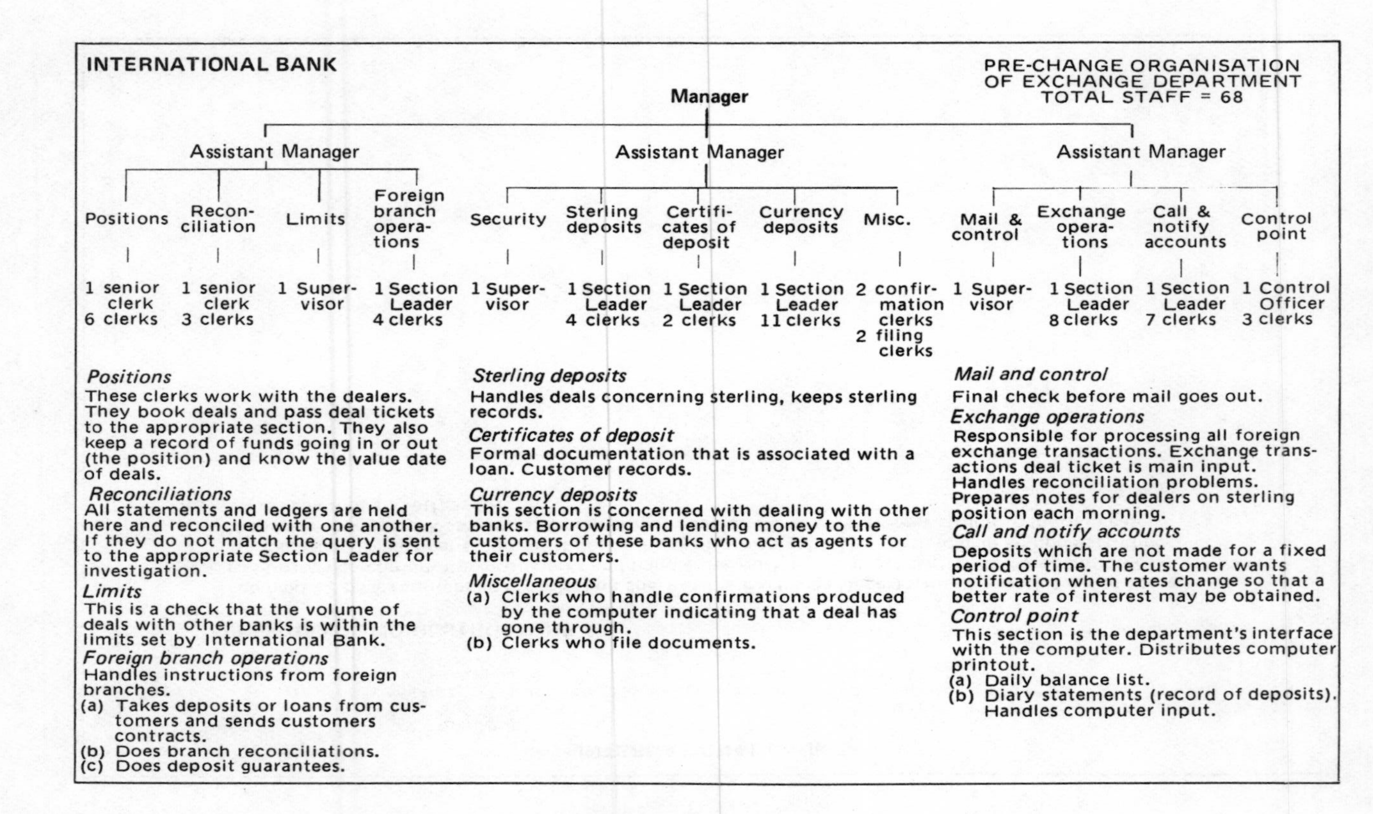

Positions
These clerks work with the dealers. They book deals and pass deal tickets to the appropriate section. They also keep a record of funds going in or out (the position) and know the value date of deals.

Reconciliations
All statements and ledgers are held here and reconciled with one another. If they do not match the query is sent to the appropriate Section Leader for investigation.

Limits
This is a check that the volume of deals with other banks is within the limits set by International Bank.

Foreign branch operations
Handles instructions from foreign branches.
(a) Takes deposits or loans from customers and sends customers contracts.
(b) Does branch reconciliations.
(c) Does deposit guarantees.

Sterling deposits
Handles deals concerning sterling, keeps sterling records.

Certificates of deposit
Formal documentation that is associated with a loan. Customer records.

Currency deposits
This section is concerned with dealing with other banks. Borrowing and lending money to the customers of these banks who act as agents for their customers.

Miscellaneous
(a) Clerks who handle confirmations produced by the computer indicating that a deal has gone through.
(b) Clerks who file documents.

Mail and control
Final check before mail goes out.

Exchange operations
Responsible for processing all foreign exchange transactions. Exchange transactions deal ticket is main input. Handles reconciliation problems. Prepares notes for dealers on sterling position each morning.

Call and notify accounts
Deposits which are not made for a fixed period of time. The customer wants notification when rates change so that a better rate of interest may be obtained.

Control point
This section is the department's interface with the computer. Distributes computer printout.
(a) Daily balance list.
(b) Diary statements (record of deposits). Handles computer input.

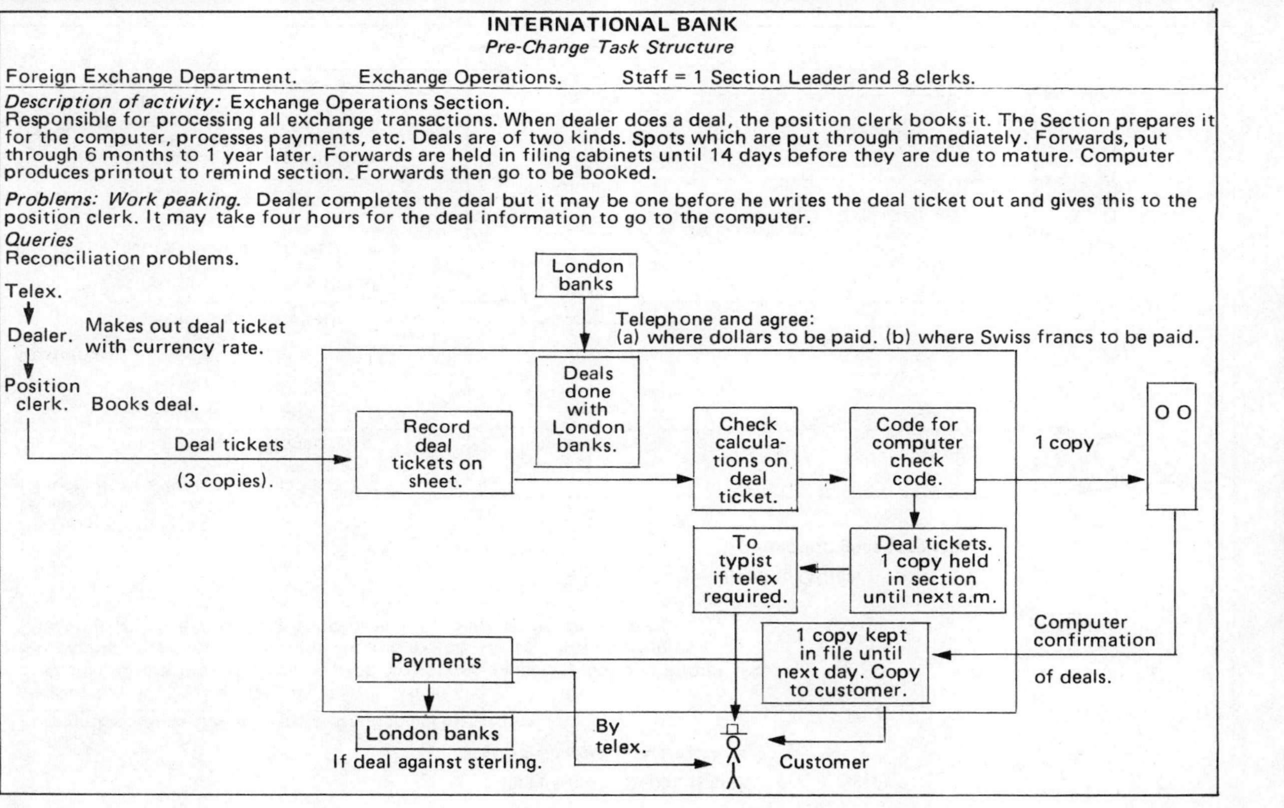
INTERNATIONAL BANK
Pre-Change Task Structure
Foreign Exchange Department.
Exchange Operations.
Staff = 1 Section Leader and 8 clerks.
Description of activity: Exchange Operations Section.
Responsible for processing all exchange transactions. When dealer does a deal, the position clerk books it. The Section prepares it for the computer, processes payments, etc. Deals are of two kinds. Spots which are put through immediately. Forwards, put through 6 months to 1 year later. Forwards are held in filing cabinets until 14 days before they are due to mature. Computer produces printout to remind section. Forwards then go to be booked.
Problems: Work peaking. Dealer completes the deal but it may be one before he writes the deal ticket out and gives this to the position clerk. It may take four hours for the deal information to go to the computer.
Queries
Reconciliation problems.
Telex.
Dealer.
Makes out deal ticket with currency rate.
Position clerk.
Books deal.
Deal tickets
(3 copies).
London banks
Telephone and agree:
(a) where dollars to be paid. (b) where Swiss francs to be paid.
Deals done with London banks.
Record deal tickets on sheet.
Check calcula-tions on deal ticket.
Code for computer check code.
1 copy
O O
Deal tickets. 1 copy held in section until next a.m.
To typist if telex required.
1 copy kept in file until next day. Copy to customer.
Computer confirmation
of deals.
Payments
London banks
If deal against sterling.
By telex.
Customer

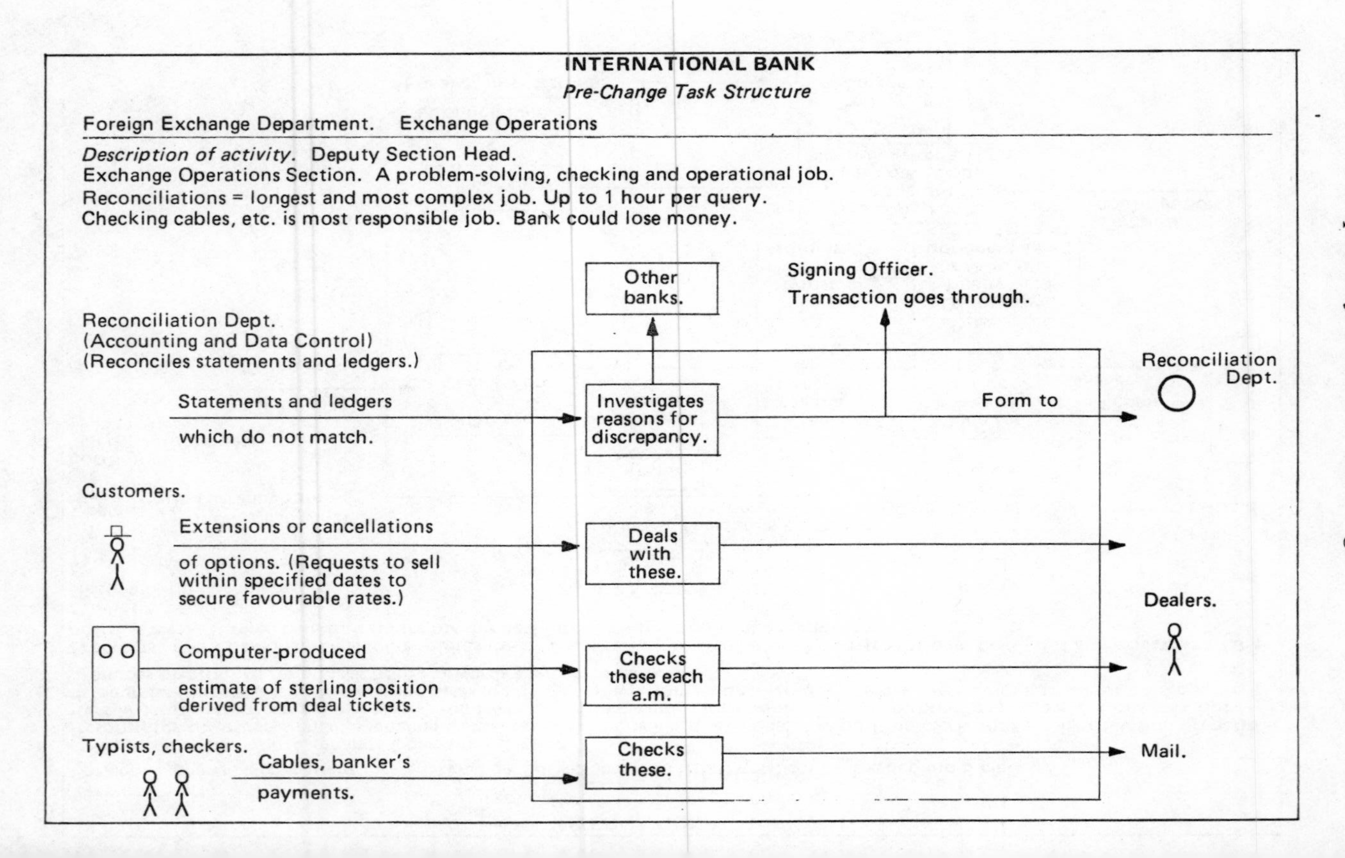
INTERNATIONAL BANK
Pre-Change Task Structure
Foreign Exchange Department. Exchange Operations
Description of activity. Deputy Section Head.
Exchange Operations Section. A problem-solving, checking and operational job.
Reconciliations = longest and most complex job. Up to 1 hour per query.
Checking cables, etc. is most responsible job. Bank could lose money.
Other banks.
Signing Officer.
Transaction goes through.
Reconciliation Dept.
(Accounting and Data Control)
(Reconciles statements and ledgers.)
Statements and ledgers
which do not match.
Investigates reasons for discrepancy.
Form to
Reconciliation Dept.
Customers.
Extensions or cancellations
of options. (Requests to sell within specified dates to secure favourable rates.)
Deals with these.
Dealers.
Computer-produced
estimate of sterling position derived from deal tickets.
Checks these each a.m.
Typists, checkers.
Cables, banker's
payments.
Checks these.
Mail.

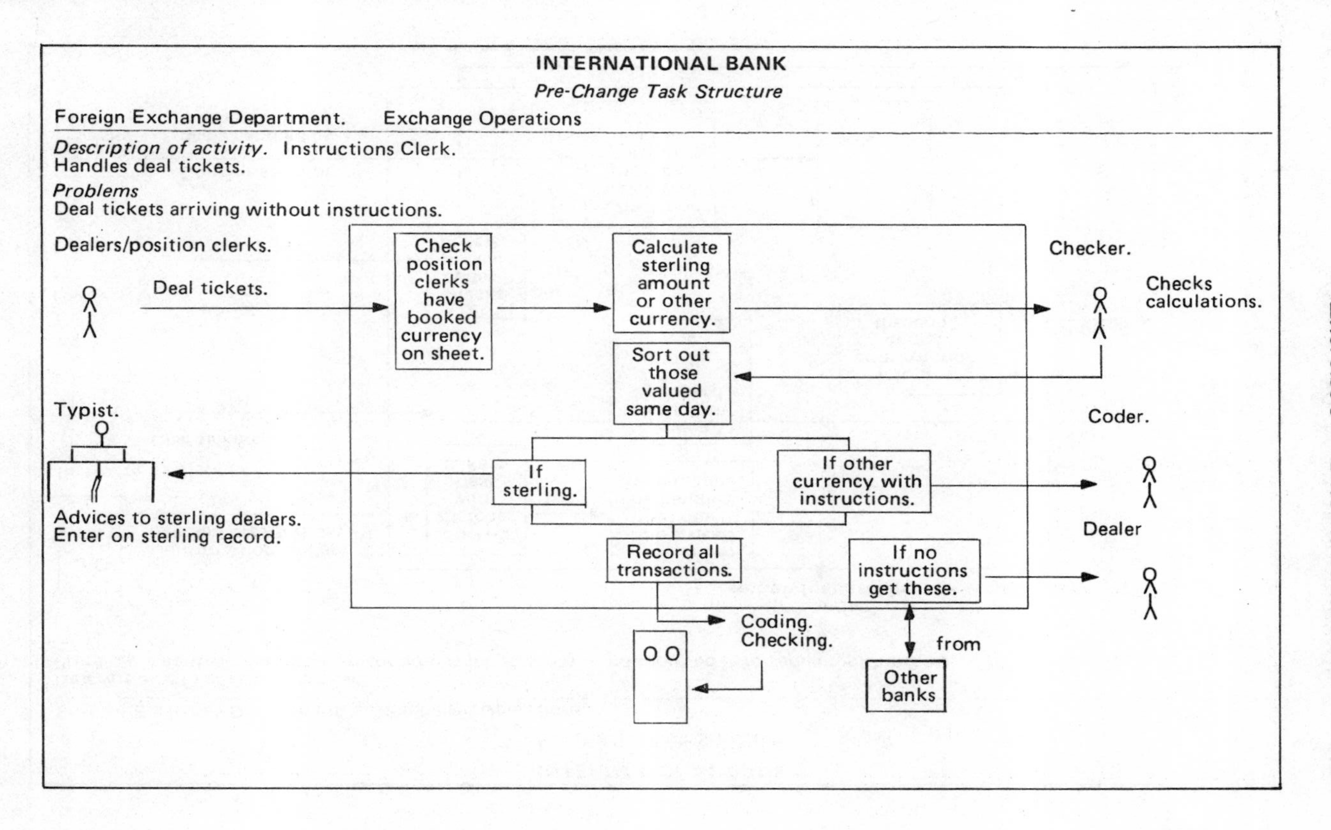

INTERNATIONAL BANK
Pre-Change Task Structure
Foreign Exchange Department. Exchange Operations
Description of activity. Instructions Clerk.
Handles deal tickets.
Problems
Deal tickets arriving without instructions.
Dealers/position clerks.
Deal tickets.
Check position clerks have booked currency on sheet.
Calculate sterling amount or other currency.
Checker.
Checks calculations.
Sort out those valued same day.
Typist.
If sterling.
If other currency with instructions.
Coder.
Advices to sterling dealers.
Enter on sterling record.
Dealer
Record all transactions.
If no instructions get these.
Coding.
Checking.
from
Other banks

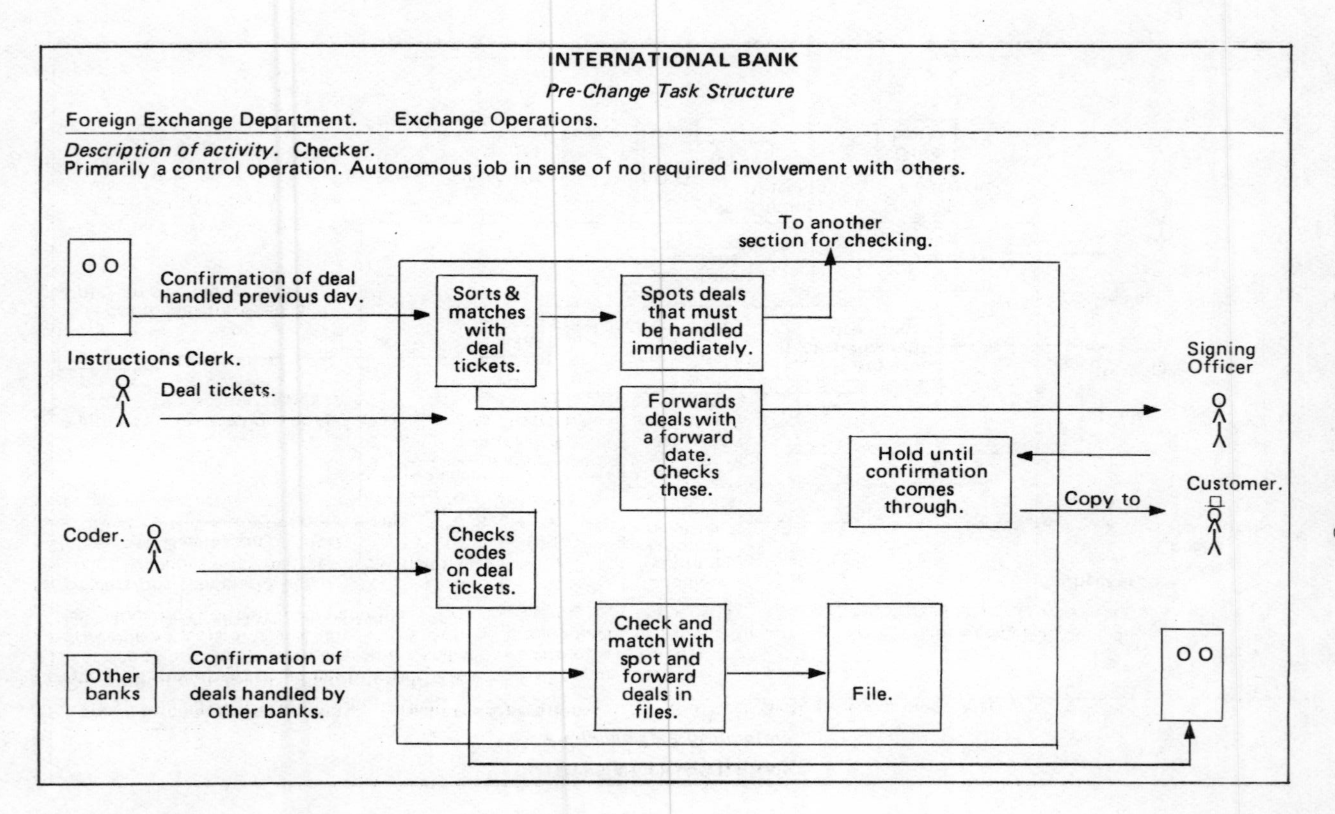
INTERNATIONAL BANK
Pre-Change Task Structure
Foreign Exchange Department. Exchange Operations.
Description of activity. Checker.
Primarily a control operation. Autonomous job in sense of no required involvement with others.
O O
Confirmation of deal handled previous day.
Instructions Clerk.
Deal tickets.
Coder.
Other banks
Confirmation of deals handled by other banks.
Sorts & matches with deal tickets.
Spots deals that must be handled immediately.
To another section for checking.
Forwards deals with a forward date. Checks these.
Signing Officer
Hold until confirmation comes through.
Copy to
Customer.
Checks codes on deal tickets.
Check and match with spot and forward deals in files.
File.
O O

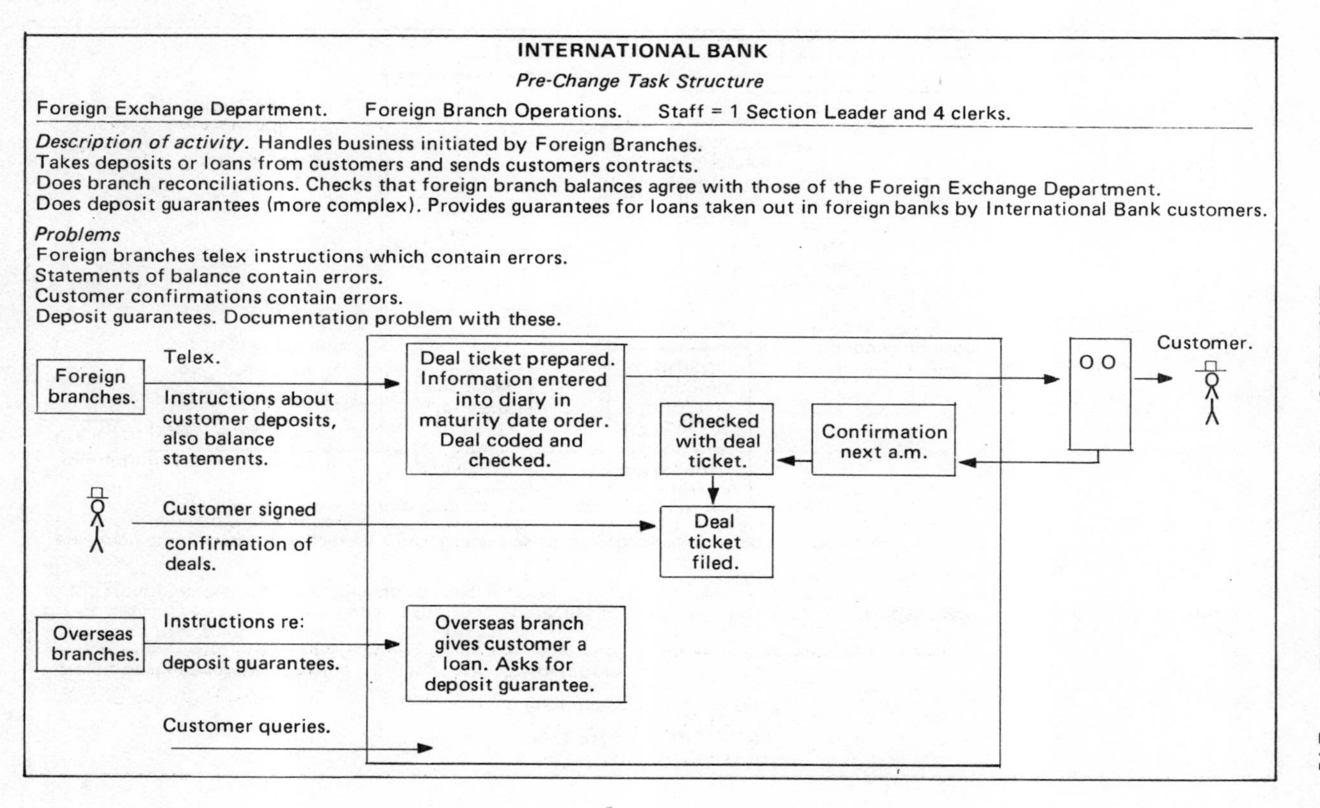
INTERNATIONAL BANK
Pre-Change Task Structure
Foreign Exchange Department. Foreign Branch Operations. Staff = 1 Section Leader and 4 clerks.
Description of activity. Handles business initiated by Foreign Branches.
Takes deposits or loans from customers and sends customers contracts.
Does branch reconciliations. Checks that foreign branch balances agree with those of the Foreign Exchange Department.
Does deposit guarantees (more complex). Provides guarantees for loans taken out in foreign banks by International Bank customers.
Problems
Foreign branches telex instructions which contain errors.
Statements of balance contain errors.
Customer confirmations contain errors.
Deposit guarantees. Documentation problem with these.
Foreign branches.
Telex.
Instructions about customer deposits, also balance statements.
Deal ticket prepared. Information entered into diary in maturity date order. Deal coded and checked.
O O
Customer.
Checked with deal ticket.
Confirmation next a.m.
Customer signed
confirmation of deals.
Deal ticket filed.
Overseas branches.
Instructions re:
deposit guarantees.
Overseas branch gives customer a loan. Asks for deposit guarantee.
Customer queries.

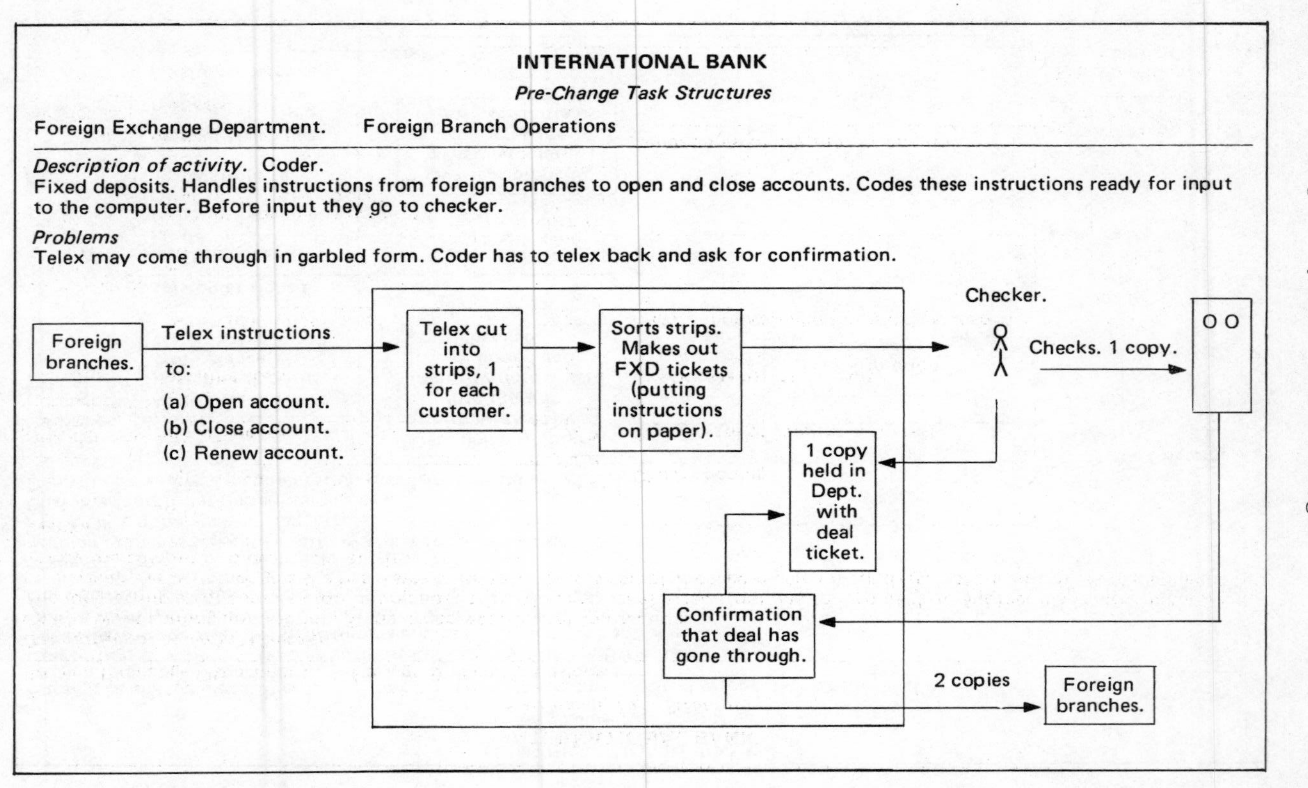
INTERNATIONAL BANK
Pre-Change Task Structures
Foreign Exchange Department.
Foreign Branch Operations
Description of activity. Coder.
Fixed deposits. Handles instructions from foreign branches to open and close accounts. Codes these instructions ready for input to the computer. Before input they go to checker.
Problems
Telex may come through in garbled form. Coder has to telex back and ask for confirmation.
Foreign branches.
Telex instructions
to:
(a) Open account.
(b) Close account.
(c) Renew account.
Telex cut into strips, 1 for each customer.
Sorts strips. Makes out FXD tickets (putting instructions on paper).
Checker.
Checks. 1 copy.
O O
1 copy held in Dept. with deal ticket.
Confirmation that deal has gone through.
2 copies
Foreign branches.

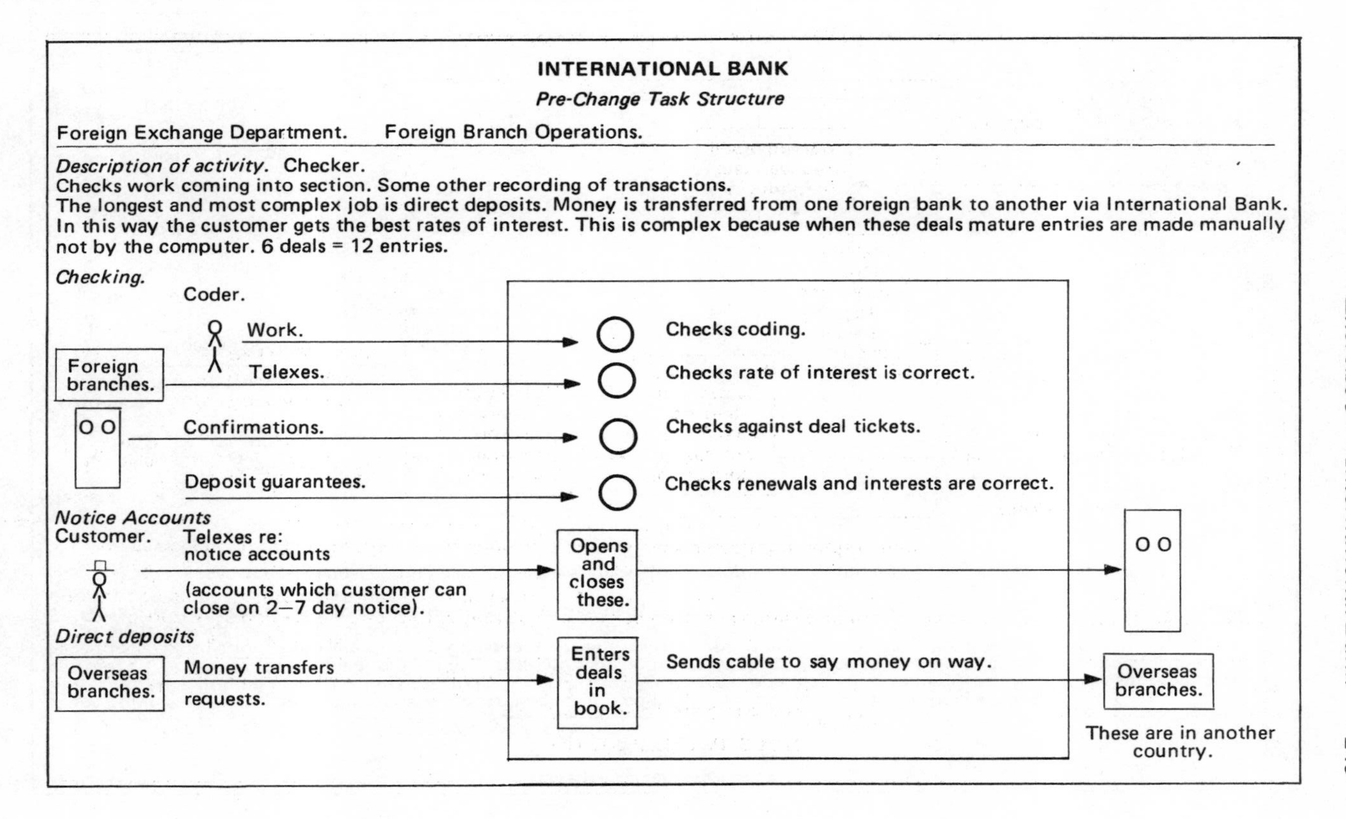
INTERNATIONAL BANK
Pre-Change Task Structure
Foreign Exchange Department. Foreign Branch Operations.
Description of activity. Checker.
Checks work coming into section. Some other recording of transactions.
The longest and most complex job is direct deposits. Money is transferred from one foreign bank to another via International Bank.
In this way the customer gets the best rates of interest. This is complex because when these deals mature entries are made manually
not by the computer. 6 deals = 12 entries.
Checking.
Coder.
Work.
Foreign branches.
Telexes.
Confirmations.
Deposit guarantees.
Checks coding.
Checks rate of interest is correct.
Checks against deal tickets.
Checks renewals and interests are correct.
Notice Accounts
Customer.
Telexes re: notice accounts
(accounts which customer can close on 2–7 day notice).
Opens and closes these.
Direct deposits
Overseas branches.
Money transfers
requests.
Enters deals in book.
Sends cable to say money on way.
Overseas branches.
These are in another country.

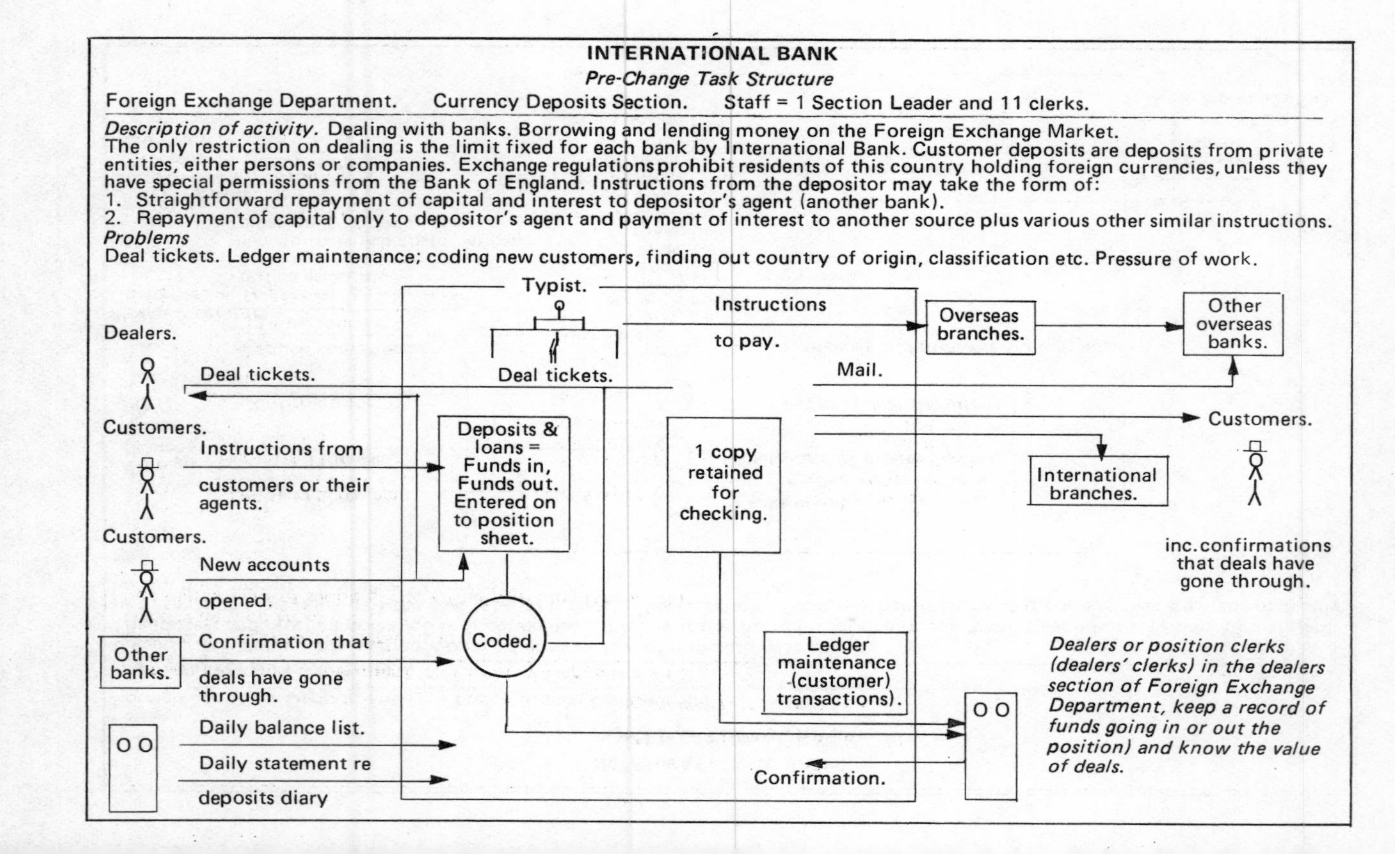
INTERNATIONAL BANK
Pre-Change Task Structure
Foreign Exchange Department. Currency Deposits Section. Staff = 1 Section Leader and 11 clerks.
Description of activity. Dealing with banks. Borrowing and lending money on the Foreign Exchange Market.
The only restriction on dealing is the limit fixed for each bank by International Bank. Customer deposits are deposits from private entities, either persons or companies. Exchange regulations prohibit residents of this country holding foreign currencies, unless they have special permissions from the Bank of England. Instructions from the depositor may take the form of:
1. Straightforward repayment of capital and interest to depositor's agent (another bank).
2. Repayment of capital only to depositor's agent and payment of interest to another source plus various other similar instructions.
Problems
Deal tickets. Ledger maintenance; coding new customers, finding out country of origin, classification etc. Pressure of work.
Typist.
Dealers.
Deal tickets.
Deal tickets.
Instructions
to pay.
Overseas branches.
Other overseas banks.
Mail.
Customers.
Instructions from
customers or their agents.
Deposits & loans = Funds in, Funds out. Entered on to position sheet.
1 copy retained for checking.
International branches.
Customers.
New accounts
opened.
inc. confirmations that deals have gone through.
Other banks.
Confirmation that
deals have gone through.
Coded.
Ledger maintenance (customer) transactions).
Dealers or position clerks (dealers' clerks) in the dealers section of Foreign Exchange Department, keep a record of funds going in or out the position) and know the value of deals.
Daily balance list.
Daily statement re
deposits diary
Confirmation.

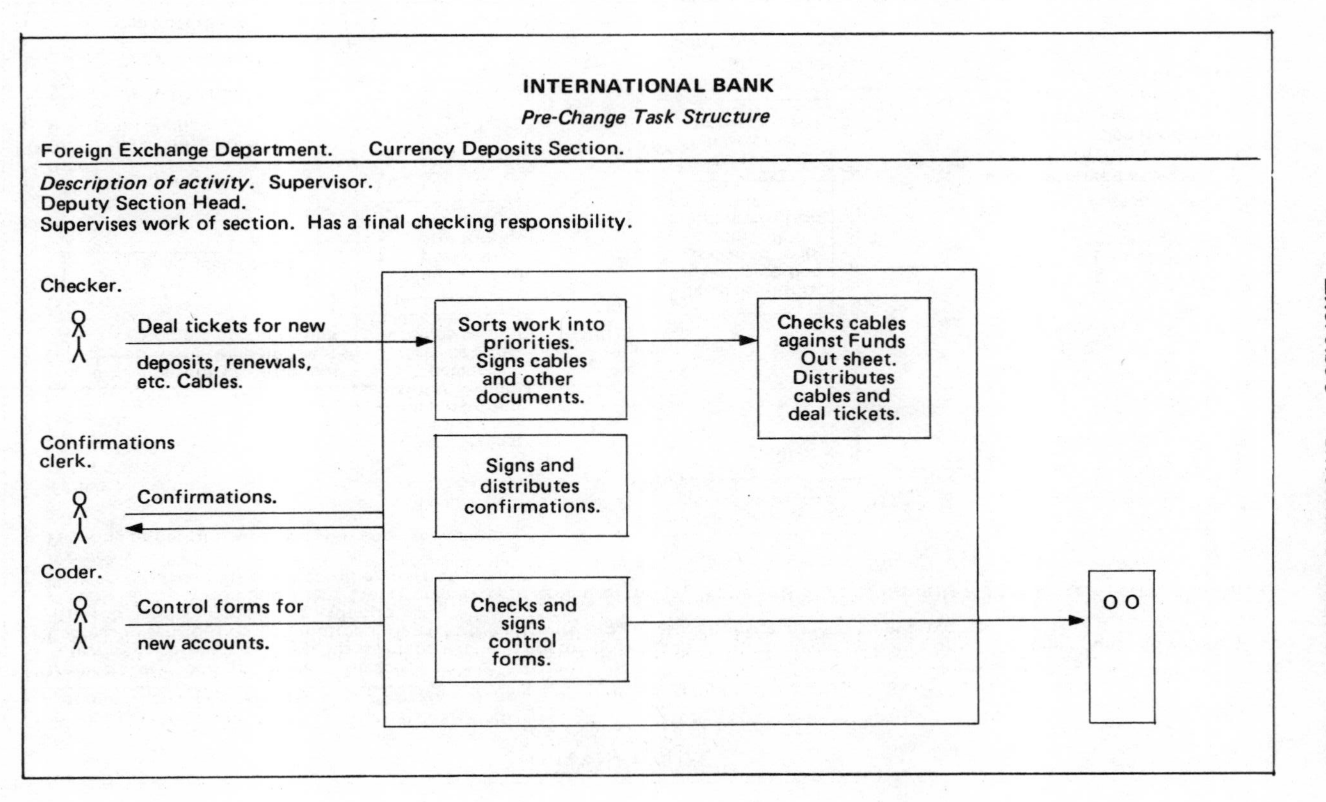
INTERNATIONAL BANK
Pre-Change Task Structure
Foreign Exchange Department. Currency Deposits Section.
Description of activity. Supervisor.
Deputy Section Head.
Supervises work of section. Has a final checking responsibility.
Checker.
Deal tickets for new deposits, renewals, etc. Cables.
Confirmations clerk.
Confirmations.
Coder.
Control forms for new accounts.
Sorts work into priorities. Signs cables and other documents.
Checks cables against Funds Out sheet. Distributes cables and deal tickets.
Signs and distributes confirmations.
Checks and signs control forms.
O O

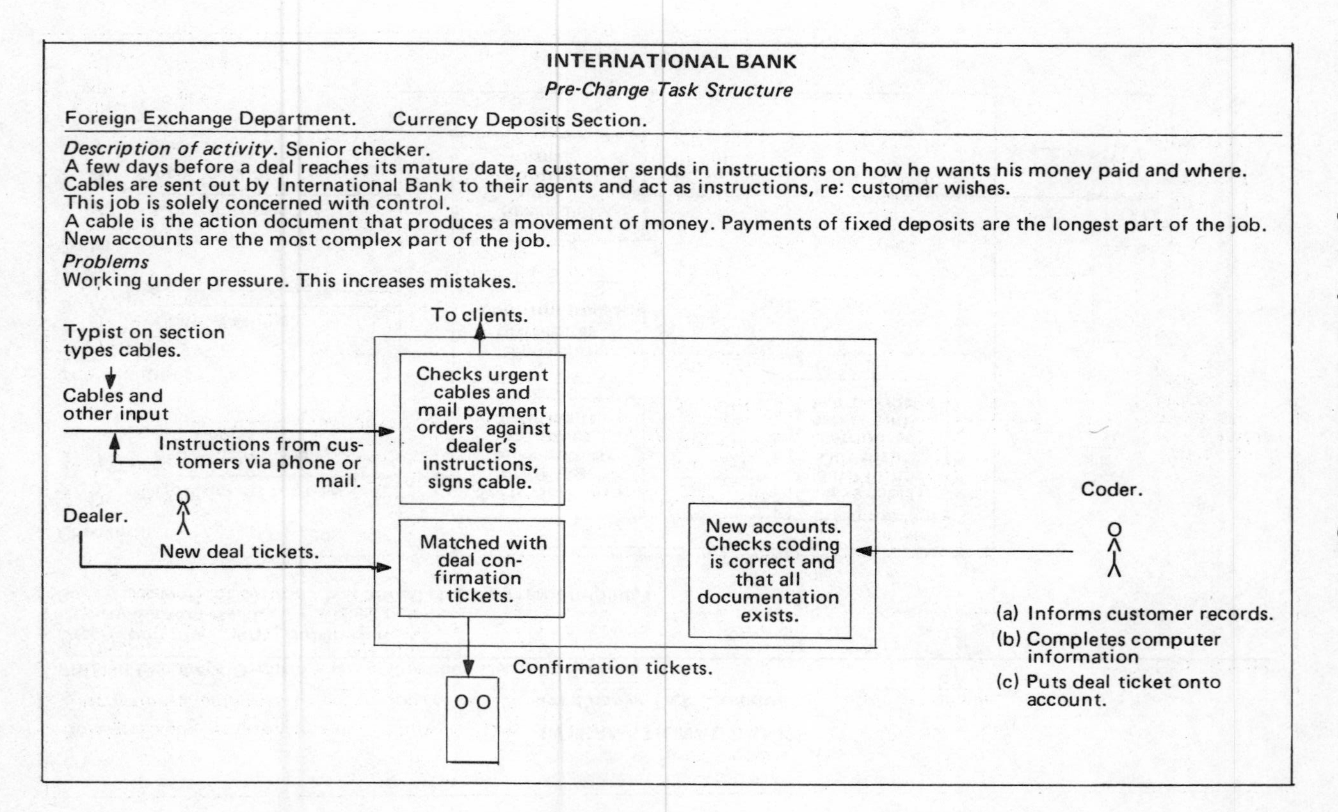
INTERNATIONAL BANK
Pre-Change Task Structure
Foreign Exchange Department. Currency Deposits Section.
Description of activity. Senior checker.
A few days before a deal reaches its mature date, a customer sends in instructions on how he wants his money paid and where.
Cables are sent out by International Bank to their agents and act as instructions, re: customer wishes.
This job is solely concerned with control.
A cable is the action document that produces a movement of money. Payments of fixed deposits are the longest part of the job.
New accounts are the most complex part of the job.
Problems
Working under pressure. This increases mistakes.
To clients.
Typist on section types cables.
Cables and other input
Instructions from customers via phone or mail.
Checks urgent cables and mail payment orders against dealer's instructions, signs cable.
Dealer.
New deal tickets.
Matched with deal confirmation tickets.
New accounts. Checks coding is correct and that all documentation exists.
Coder.
(a) Informs customer records.
(b) Completes computer information
(c) Puts deal ticket onto account.
Confirmation tickets.
O O

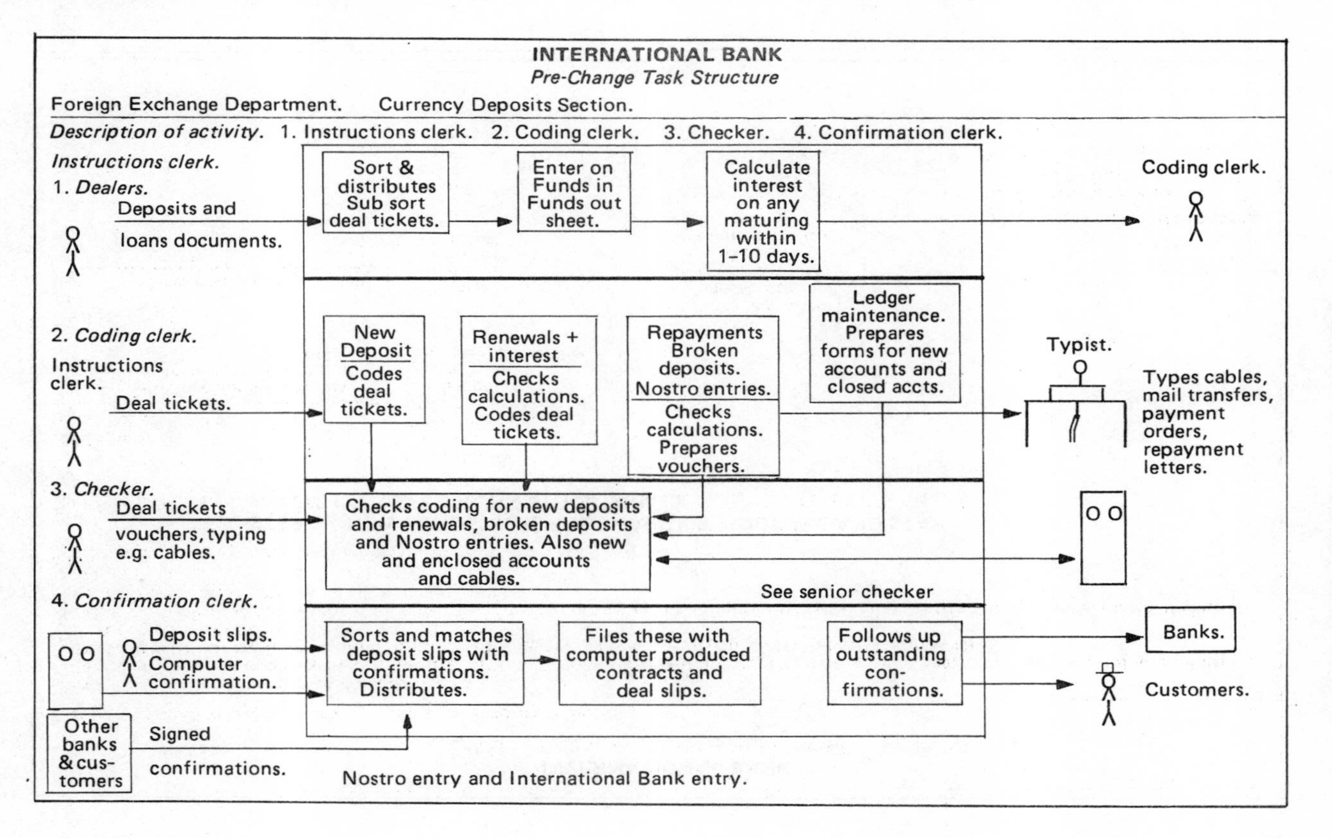
INTERNATIONAL BANK
Pre-Change Task Structure
Foreign Exchange Department.
Currency Deposits Section.
Description of activity.
1. Instructions clerk.
2. Coding clerk.
3. Checker.
4. Confirmation clerk.
Instructions clerk.
1. Dealers.
Deposits and
loans documents.
Sort & distributes Sub sort deal tickets.
Enter on Funds in Funds out sheet.
Calculate interest on any maturing within 1–10 days.
Coding clerk.
2. Coding clerk.
Instructions clerk.
Deal tickets.
New Deposit
Codes deal tickets.
Renewals + interest
Checks calculations. Codes deal tickets.
Repayments Broken deposits. Nostro entries.
Checks calculations. Prepares vouchers.
Ledger maintenance. Prepares forms for new accounts and closed accts.
Typist.
Types cables, mail transfers, payment orders, repayment letters.
3. Checker.
Deal tickets
vouchers, typing e.g. cables.
Checks coding for new deposits and renewals, broken deposits and Nostro entries. Also new and enclosed accounts and cables.
See senior checker
4. Confirmation clerk.
Deposit slips.
Computer confirmation.
Sorts and matches deposit slips with confirmations. Distributes.
Files these with computer produced contracts and deal slips.
Follows up outstanding con-firmations.
Banks.
Customers.
Other banks & cus-tomers
Signed
confirmations.
Nostro entry and International Bank entry.

INTERNATIONAL BANK

STEP 3 NOW FORGET THE PRE-CHANGE STRUCTURE AND START TO DEVELOP A NEW, IMPROVED WORK STRUCTURE.

INTERNATIONAL BANK

Your new system should include the operations in the diagram set out below:

TECHNICAL SYSTEM		TASK SYSTEM
Input. The equipment used to enter data into the computer—different types of equipment can create very different jobs for staff. Communication to and from branches of International Bank can also be considered here.	Input	*Input.* Jobs dealing with data related to deals. 1. Customer's name, rate, date of deal, currency amount. 2. Information on new customers for master record. 3. Information used to assist the preparation of advices to customers and other banks.
Computer. The computer can do most of the work needed to process deals, update ledgers and identify errors. Work can be done either in batches or in real-time (immediate update of records).	Computer update	*Computer.* Dealing with updating of agent's and customers' accounts and bank ledgers.
Output. The computer can produce information on to a terminal, paper printout, or microfilm. Output to be dealt with by staff will include: 1. Confirmation that deals have gone through. This information is used to inform customers and other banks and for filing in bank's records. 2. Advices and bankers' payments. 3. Daily balance list. 4. Diary statements (record of deposits). 5. Currency positions. 6. Information for management.	Output	*Output jobs.* Dealing with the documents and information produced by the computer (see opposite). In particular ensuring that these are accurate. Obtaining information with which to answer queries. This will be obtained both from the computer and other sources such as customers and banks.

Flow of work through office.

INTERNATIONAL BANK

STEP 3 SETTING OUT ALTERNATIVE SOLUTIONS

Technical solutions

It may be easier to break the technical system down into 3 stages:

1. *Input* The method by which data gets into the computer.
2. *Updates* Whether accounts are processed straight away (on-line/real time), or later in batches – and how often the batches are run. A combination of real time and batch processing is also possible.
3. *Output* Whether the information produced by the computer is printed out at once on-line through a terminal, or printed out on paper or microfilm, and what form the printouts should take.

Set out on the next 3 pages several different technical solutions which you think might be appropriate for the new computer system, bearing in mind the objectives which have been set. Four or five alternatives would be a reasonable number to consider, though of course you may consider more or fewer alternatives if you wish. Some alternatives might differ only in one aspect, e.g. method of input or the form of the printout, so it may help to group these together under a broad heading with the differences as subheadings. For each solution or group of solutions, set out the technical *and* social advantages and disadvantages which you think it may have.

INTERNATIONAL BANK

Description of technical solution	*Advantages*	*Disadvantages*
(a) *Input* (b) *Update*	*Technical and business advantages*	*Technical and business disadvantages*
(c) *Output*	*Job-satisfaction advantages*	*Job-satisfaction disadvantages*
(a) *Input* (b) *Update*	*Technical and business advantages*	*Technical and business disadvantages*
(c) *Output*	*Job-satisfaction advantages*	*Job-satisfaction disadvantages*

INTERNATIONAL BANK		
Description of technical solution	Advantages	Disadvantages
(a) Input	Technical and business advantages	Technical and business disadvantages
(b) Update	Job-satisfaction advantages	Job-satisfaction disadvantages
(c) Output		
(a) Input	Technical and business advantages	Technical and business disadvantages
(b) Update	Job-satisfaction advantages	Job-satisfaction disadvantages
(c) Output		

INTERNATIONAL BANK

Description of technical solution	*Advantages*	*Disadvantages*
(a) *Input* (b) *Update*	*Technical and business advantages*	*Technical and business disadvantages*
(c) *Output*	*Job-satisfaction advantages*	*Job-satisfaction disadvantages*
(a) *Input* (b) *Update*	*Technical and business advantages*	*Technical and business disadvantages*
(c) *Output*	*Job-satisfaction advantages*	*Job-satisfaction disadvantages*

INTERNATIONAL BANK

Now do a preliminary evaluation of the alternative *technical* solutions you have put forward.

For each solution ask yourself:

1. Does it achieve the priority technical and business objectives which have been specified in Step 2 (page 233).
   Does it achieve all or some of the non-priority technical and business objectives specified in Step 1 (pages 221-2).
2. Is it limited in any way by the constraints identified earlier?
3. Are the available resources which were identified earlier adequate to achieve this solution?

You can now draw up a short-list of technical solutions which still seem reasonable after they have been examined in this way.

Try and keep your short-list down to three or four solutions. Write out your short-list on the next page.

INTERNATIONAL BANK

Technical solutions short-list

Give brief description

1.

2.

3.

4.

INTERNATIONAL BANK

REMEMBER THERE ARE NO COMPLETELY RIGHT OR WRONG TECHNICAL SOLUTIONS. OUR SUGGESTIONS ARE ON THE FOLLOWING PAGES.

INTERNATIONAL BANK
Technical Solution

Description of technical solution	*Advantages*	*Disadvantages*
T.1 (a) *Input* Key punch input onto paper or magnetic tape. (b) *Update* Batch processing	*Technical and business advantages* 1. Cheapest and easiest method to update when enough information is ready. 2. Easy to recover in case of breakdown.	*Technical and business disadvantages* 1. Needs 12 weeks' training, therefore very dependent on operators. Could be difficult e.g. in flu epidemic. 2. Time delay before rejects are dealt with because batch processing system holds orders up. 3. Paper printout tends to be bulky.
(c) *Output* Paper printouts for all output, produced the following day.	*Job-satisfaction advantages* 1. Similar to using typewriter so it is easier for girls who have been typists. 2. Paper printout is fairly easy to read, if it is well set out.	*Job-satisfaction disadvantages* 1. Girls do not see what they are punching—can be frustrating. 2. Batch processing system can cause problems for staff dealing with rejects—they are delayed if computer breaks down and then have to cope with build-up of work. 3. Paper printout is heavy to handle and store.
T2. (a) *Input* Visual display units (VDUs) on-line to small computer—which does some checking and creates a magnetic tape.	*Technical and business advantages* 1. Flexible system—mistakes can be corrected immediately before they go on to magnetic tape. 2. Fast system at input stage. 3. Needs only 5 weeks' training, therefore not as dependent on expertise.	*Technical and business disadvantages* 1. Need a larger computer, or an additional small computer to run the VDUs. 2. VDUs are relatively costly. 3. Batch processing may not meet the needs of the dealer.
(b) *Update* Batch processing from magnetic tape. (c) *Output* Paper printouts for all output produced the following day.	*Job-satisfaction advantages* 1. Reasonably satisfying job for the VDU operators, as they are interacting with the computer via VDUs. 2. Operators can understand what the computer displays on the screen—no special languages used or holes in tape, etc.	*Job-satisfaction disadvantages* 1. Can still be fairly repetitive job when dealing with a large number of documents which are all the same. 2. VDU operation can sometimes cause eyestrain and headaches. 3. Batch processing may not meet the needs of the dealer.

INTERNATIONAL BANK
Technical Solution

Description of technical solution	*Advantages*	*Disadvantages*
T3. (a) *Input* VDU on-line data entry with VDU enquiry facility. (b) *Update* Batch processing. (c) *Output* Paper printouts next day. Historical information on customers and agents can be obtained via VDUs.	*Technical and business advantages* As in T2. 1. Clerks can get immediate access to customers' information as a help to solving problems and answering enquiries. 2. More information becomes available. *Job-satisfaction advantages* 1. More satisfying job for VDU operators than T2. Two-way interaction with computer is now possible, less repetitive. 2. Operator can understand what the computer displays on the screen as in T2.	*Technical and business disadvantages* As in T2. 1. Information obtained via VDUs will be up to 24 hours out of date. 2. Slow compared with real-time. 3. Batch processing may not meet the needs of the dealer. *Job-satisfaction disadvantages* 1. VDU operation can sometimes cause eyestrain and headaches.
T4. (a) *Input* Via VDU terminals, on-line direct to computer, VDU enquiry facility. (b) *Update* Real time—deal records and accounts updated at once. (c) *Output* Paper printouts, available immediately if required. Up-to-date information available immediately via VDUs.	*Technical and business advantages* 1. Very fast system, including correction of mistakes and deal record and accounts always up to date. 2. Immediate production of deal contracts. 3. Big saving in paper. 4. More information becomes available. *Job-satisfaction advantages* 1. Can be satisfying to interact directly with the computer—asking questions and receiving replies. 2. Information for solving problems and answering queries is available right away. 3. Speeds up confirmations of deals with customers.	*Technical and business disadvantages* 1. Advanced system—expensive to develop. Difficult to justify the costs. 2. Need more expensive equipment to update immediately. 3. More chance of breakdowns. 4. A technically complex system. *Job-satisfaction disadvantages* 1. See T3.

INTERNATIONAL BANK
Technical Solution

Description of technical solution	*Advantages*	*Disadvantages*
T5. (a) *Input* Via VDU terminals, on-line direct to computer. VDU enquiry facility. (b) *Update* Real time for update of deals. Batch processing of information for bank.	*Technical and business advantages* 1. Allows more time for system development if later move to total real-time. 2. Meets dealer's requirements. Exchange Department has less need of completely up to date information. See T4.	*Technical and business disadvantages* 1. Slower system in Exchange Department area. See T4.
(c) Paper printouts re. deal information available immediately if required. Up to date information on deals available via VDUs. Bank information via printouts and VDUs is historical.	*Job-satisfaction advantages* 1. Can be satisfying to interact with computer—asking questions and receiving replies. 2. Information for solving some problems and queries is available immediately and up to date. 3. Other information is available immediately, although not up to date.	*Job-satisfaction disadvantages* See T3.
T6. By package system for money market.	*Technical and business advantages* 1. Lower costs.	*Technical and business disadvantages* 1. Would not meet all of International Bank's requirements. 2. There could be maintenance problems. If the system goes wrong it could be difficult to correct.
	Job-satisfaction advantages These would be unknown until the system was installed. Social constraints may be already built into the system.	*Job-satisfaction disadvantages* See job-satisfaction advantages.

INTERNATIONAL BANK

Technical solutions short-list

T2. Visual display units on-line for input.
Batch processing update.
Paper printout with historical information.

T3. Visual display units on-line for input.
VDU enquiry facility.
Batch processing update.
Output via paper printout and VDUs.
All information historical.

T4. Visual display units on-line for input.
VDU enquiry facility.
Real-time update for all operations.
Immediate paper printout if required.
Up to date information via VDUs.

T5. Visual display units on-line for input.
VDU enquiry facility.
Real-time update for deals.
Batch processing update for bank information.
Output via paper printout and VDUs.
Deal information up to date.
Bank information historical.

T1. Was rejected as too slow, particularly with regard to error correction.

T6. Was rejected as being unlikely to meet all of International Bank's requirements.

INTERNATIONAL BANK

STEP 3 SOCIAL SOLUTIONS

Background information for social solutions

You have already examined the 'fit' between employee needs and experience on our five variables: (1) Use of skills and knowledge; (2) Psychological needs; (3) The methods used to ensure efficiency, including information, material, salary and work controls; (4) The need to work for a bank whose ethics and values fit those of the employee; (5) The need for a task structure which meets needs for variety, autonomy, etc. The bank too has requirements on all these variables and these influence its policies. International Bank's policies are set out below.

1. *Knowledge policy*

The bank tries to recruit people of good intelligence preferably with 5 or more 'O' levels. They do not always find it easy to do this and, once recruited, staff are difficult to keep. Because of this the bank employs many part-time clerks. A number of jobs are very simple and sufficient knowledge can be gained to do these in 2–3 weeks. At the lower levels of the grading structure there is a clear demarcation between jobs, so that each person only knows about her own job. The nature of banking means that a great many controls are included in the work system.

2. *Policies for meeting employee's psychological needs*

The management of the bank has tried to create a family feeling and is very concerned to establish good relationships between management and staff. It wishes to encourage a high level of commitment on the part of its staff and to provide a situation in which staff enjoy their work and believe that their job is worth doing. Top management recognises that the present organisation of work does not meet staff needs in these respects.

INTERNATIONAL BANK

Background information for social solutions (continued)

3. *Policies for ensuring efficiency (support/control system)*

   The standard of work is very high because the bank has a great need for accuracy. But the tight control systems which help ensure accuracy can reduce staff opportunities for using discretion and taking decisions. Supervisory support is excellent, particularly at Section Leader level where considerable help and assistance is given to junior staff.

4. *Policy on task structure*

   The work of the department is sub-divided into a number of different tasks, with staff grouped into sections to deal with these tasks. The bank requires many people who can keep at a repetitive job and achieve a good work output. The lower-level jobs require very little skill and initiative.

5. *Ethical approach*

   The bank has a very responsible attitude to its staff and wishes to provide a good working environment, job satisfaction, good relationships, satisfactory pay and conditions and good promotional opportunities. It encourages staff discussion of new policies.

INTERNATIONAL BANK

STEP 3 SETTING OUT ALTERNATIVE SOCIAL SOLUTIONS

Now proceed to set out alternative social solutions, though at this stage you may think about them independently of the technical solutions which you have just set out. You may keep input and output tasks separate or combine them together, but remember the update tasks are performed by the computer itself. It may also be helpful to think of the social system as a two-part system. The first part is the organisation of the total workflow into its various units or sections. Later on comes the design of the actual work which is undertaken in each of these sections. At present, concentrate on the total office organisation, and consider the more detailed job design in Step 6.

INTERNATIONAL BANK
Social Solution

Description of social solution	*Advantages*	*Disadvantages*
(a) *Input*	*Job-satisfaction advantages*	*Job-satisfaction disadvantages*
(b) *Output*	*Technical and business advantages*	*Technical and business disadvantages*
(a) *Input*	*Job-satisfaction advantages*	*Job-satisfaction disadvantages*
(b) *Output*	*Technical and business advantages*	*Technical and business disadvantages*

INTERNATIONAL BANK *Social Solution*		
Description of social solution	*Advantages*	*Disadvantages*
(a) *Input*	***Job-satisfaction advantages***	***Job-satisfaction disadvantages***
(b) *Output*	***Technical and business advantages***	***Technical and business disadvantages***
(a) *Input*	***Job-satisfaction advantages***	***Job-satisfaction disadvantages***
(b) *Output*	***Technical and business advantages***	***Technical and business disadvantages***

INTERNATIONAL BANK

When you have set up several alternative ways of organising the departments concerned, do a preliminary evaluation of your social solutions as you did for the technical solution as follows:

1. Do the solutions achieve priority human objectives which have been identified in Step 2 (page 234).
   Do they achieve some or all of the non-priority human objectives specified in Step 1 (pages 215-22).

2. Are any of the solutions constrained by the conditions which you identified earlier?

3. Have you got the resources available to achieve each of the solutions you have set up?

Having checked all your solutions against these three criteria, eliminate any about which you are doubtful, and draw up a short-list of the remainder, if possible no more than three or four solutions. Enter your short-list of social solutions on the right-hand side of the next page. Copy your list of preferred technical solutions which you wrote in on page 255 onto the left-hand side.

INTERNATIONAL BANK

Technical solutions short-list *Give brief description*	*Social solutions short-list* *Give brief description*
1.	1.
2.	2.
3.	3.
4.	4.

INTERNATIONAL BANK

OUR SUGGESTIONS ARE ON THE FOLLOWING PAGES.

INTERNATIONAL BANK
Social Solution

Description of social solution	*Advantages*	*Disadvantages*
S1. The present functional organisation of the office is retained. The work of a section is associated with a particular kind of task or transaction.	*Job-satisfaction advantages* Some staff like the present organisation of work and do not want it altered. This is particularly true of more senior staff who have considerable knowledge of particular transactions.	*Job-satisfaction disadvantages* Many lower-level staff find the present organisation of work dull and routine. The present functional organisation does not provide much scope for job enrichment if there is a wish for this in the future.
	Technical and business advantages Staff are familiar with this organisation of work and operate it efficiently.	*Technical and business disadvantages* This arrangement tends to be vulnerable and inflexible if some staff are ill or on holiday.
S2, The office is divided into groups which deal with *all* stages of the work associated with a particular currency,—but at lower-grade levels each clerk is responsible for one task only, e.g. data entry via VDUs.	*Job-satisfaction advantages* 1. The staff will be working in a group, even though many are only doing one task, so they can identify with that group. 2. Staff can move up the hierarchy of jobs associated with handling a particular currency. 3. The jobs will be very similar to the type of jobs in the present system, which some staff prefer.	*Job-satisfaction disadvantages* 1. Some staff may find it boring and repetitive if they are not doing one task. This is no improvement on the present arrangement under the batch computer system.
	Technical and business advantages 1. The work may go through more quickly, if each person is concentrating on one task. Accuracy may be improved.	*Technical and business disadvantages* 1. This arrangement tends to be vulnerable and inflexible if some staff are ill or on holiday. 2. New knowledge is required in order to deal with a particular currency.

INTERNATIONAL BANK
Social Solution

Description of social solution	*Advantages*	*Disadvantages*
S3. The office is divided into groups with each clerk doing *all* stages of the work associated with a particular currency. Obtaining resources such as information and materials together with problem solving is a supervisory responsibility carried out by Section Leaders and their deputies.	*Job-satisfaction advantages* 1. Staff like working in groups. This creates feelings of group identify. 2. The work is interesting as each girl has a variety of tasks. 3. A good learning environment. As one task is mastered, another can be learnt.	*Job-satisfaction disadvantages* 1. This form of work organisation would affect the grading structure at its lower levels. 2. A much longer training period would be needed for these jobs and so it would increase social costs for the bank, particularly if labour turnover continues to be high.
	Technical and business advantages 1. This arrangement provides work flexibility. If some staff are ill, others know their work. 2. The quality of work should be better if one person is responsible for a number of tasks.	*Technical and business disadvantages* 1. This arrangement could increase security problems so good controls will be needed. 2. The throughput of work may be rather slower if one person is doing a number of stages of the work.
S4. The office is divided into groups with each clerk doing all stages of the work associated with a particular currency. Clerks are able to requisition their own materials and search for necessary information. Difficult problems are allocated amongst group members. The Section Leader's role is a co-ordinating, training and personnel oriented one. The Deputy Section Leader becomes superfluous.	*Job-satisfaction advantages* 1. The work is more interesting and challenging than in S3. The more difficult problem-solving work is now undertaken by members of the group, not the supervisor. See S3 above.	*Job-satisfaction disadvantages* 1. This organisation of work would alter the existing grading structure in a major way. 2. The role of the Section Leader is greatly changed and Section Leaders may not like the new role. See S3 above.
	Technical and business advantages See S3 above.	*Technical and business disadvantages* A high level of responsibility is diffused throughout the group. Some clerks may not have the knowledge or intelligence to go with this. See S3 above.

INTERNATIONAL BANK

Drawing up the short-list of social solutions

There did not seem to be any real human advantage in keeping the present situation by using solution S1 and so S1 was excluded from the short-list.

INTERNATIONAL BANK

Technical solutions short-list		*Social solutions short-list*	
	Give brief description		*Give brief description*
T2.	Visual display units on-line for input. Batch processing update. Paper printout with historical information.	S2.	Office divided into longitudinal groups which deal with all stages of the work associated with a particular currency, but each clerk does only one task; for example, handles computer input, unless they are a Section Leader or Deputy Section Leader.
T3.	Visual display units on-line for input. VDU enquiry facility. Batch processing update. Output via paper printout and VDUs. All information historical.	S3.	Office divided into longitudinal groups which deal with all stages of the work associated with a particular currency. Each clerk does all tasks. Obtaining resources, final checking and major problem solving remains a supervisory responsibility.
T4.	Visual display units on-line for input. VDU enquiry facility. Real-time update for all operations. Immediate paper printout if required. Up-to-date information via VDUs.	S4.	Office divided into longitudinal groups which deal with all stages of the work associated with a particular currency. Each clerk does all tasks. Clerks can requisition materials, obtain necessary information, solve problems. The Section Leader's role becomes primarily a co-ordinating, training, personnel role with some checking and signing responsibilities. The Deputy Section Leader's role becomes unnecessary.
T5.	Visual display units on-line for input. VDU enquiry facility. Real-time update for deals. Batch processing update for bank information. Output via paper printout and VDUs. Deal information up to date. Bank information historical.		

INTERNATIONAL BANK

STEP 4 SET OUT POSSIBLE SOCIO-TECHNICAL SOLUTIONS

It is now necessary to merge your short-list of technical and social solutions.

The essential thing is to see which technical and social solutions are compatible with one another and to eliminate any technical or social solution which cannot be fitted to a compatible social or technical solution.

Take each solution in turn and compare it with all the solutions on your other short-list. Where the two solutions could be operated together, mark this combination of solutions down as a possible socio-technical solution on the next page. It may be that *all* your solutions are compatible with one another, in which case you can enter them all as possible socio-technical solutions.

Please use your social and technical solutions, not ours.

Do not rank the solutions in order of preference yet.

INTERNATIONAL BANK

Possible Socio-technical Solutions Short-list

Description	*Ranking*	*Description*	*Ranking*	*Description*	*Ranking*
1.		2.		3.	
4.		5.		6.	
7.		8.		9.	
10.		11.		12.	

INTERNATIONAL BANK

STEP 5 RANKING SOCIO-TECHNICAL SOLUTIONS

The list of socio-technical solutions which you have just drawn up must now be ranked, before the most suitable system can be chosen. Turn back to Step 3, where you set out the technical and business and job-satisfaction advantages and disadvantages of the various solutions you put forward. Consider what you wrote on those pages then try and rank the socio-technical solutions on the previous page from 1 — 12.

When you feel that you have achieved a satisfactory ranking, enter this against each socio-technical solution on the previous page. It should now be clear which socio-technical solution you consider the best, but it is necessary to check that this solution is *completely* satisfactory before accepting it as the most suitable system.

Our ranked socio-technical solutions are shown on the next page.

INTERNATIONAL BANK

Possible Socio-technical Solutions Short-list

Description	*Ranking*	*Description*	*Ranking*	*Description*	*Ranking*
1. T2. Visual display units. Batch processing. Paper printout. + S2. Longitudinal groups dealing with all stages of work. Each clerk does one special task.	12	2. T2 + S3. Longitudinal groups dealing with all stages of work. Each clerk does all tasks.. Supervisors get resources, check, solve major problems.	8	3. T2 + S4. Longitudinal groups dealing with all stages of work. Each clerk does all tasks, including getting resources, checking, solving problems.	7
4. T3. Visual display units with enquiry facility. Batch processing. Paper printout. + S2	11	5. T3 + S3	5	6. T3 + S4	6
7. T4. Visual display units with enquiry facility. Real time processing. Paper printout. + S2	10	8. T4 + S3	3	9. T4 + S4	4
10. T5. Visual display units with enquiry facility. Real time for deals. Batch for bank information. Paper printout. + S2	9	11. T5 + S3	2	12. T5 + S4	1

INTERNATIONAL BANK

CHECK YOUR CHOSEN SOCIO-TECHNICAL SOLUTION

Although you have already checked the technical and social solutions *separately*, it is now necessary to make sure that your combined socio-technical solution still meets the criteria laid down earlier. Turn back to Steps 1 and 2 and check that your chosen socio-technical solution meets the conditions set out there, e.g.

1. Does this solution meet *both* technical and business objectives and social needs? (see pages 221-2).
2. Are sufficient resources available to achieve *both* the technical and social aspects of your chosen solution?
3. Do any of the constraints set out earlier make your chosen solution impossible?
4. Does this solution meet some of the other technical and social objectives which were set out on pages 231 and 232, but not made priority objectives.

If you are satisfied that your chosen socio-technical solution is still viable proceed to Step 6 and prepare a detailed work design. If you find that your chosen solution has not met the conditions above, return to your short-list of socio-technical solutions, take the solution ranked next and check this against these criteria. Carry on until you find a socio-technical solution which meets the criteria.

INTERNATIONAL BANK

STEP 6 PREPARE A DETAILED WORK DESIGN FROM YOUR CHOSEN SOCIO-TECHNICAL SOLUTION

Prepare a list and a description of all the tasks which will be created for people to do if you implement your chosen socio-technical solution. A simple flow chart would be helpful. Describe how you propose to allocate these tasks amongst the different work groups. How many work groups will there be?

Now rank these tasks in order of simplicity, specifying which are to be undertaken by grades –1, –2 and –3 clerks and which by Section Leaders and Deputy Section Leaders.

Now check that this arrangement of tasks will create jobs which are as interesting and satisfying as possible.

1. Is each clerk able to use a variety of different skills and different levels of skill?
2. Can each clerk easily identify the targets he/she will have to achieve?
3. Are there feedback mechanisms to tell them if they are achieving these targets?
4. Are there opportunities in each job for the clerk to make choices and use discretion?
5. Is the cycle time of the different tasks or task mix long enough to avoid a feeling of repetitive work but short enough to allow the employee to feel he is making progress with his work?
6. Can each clerk feel he is making a significant contribution to the work of the bank?

If you consider that the jobs you have created are as satisfying as they could be, while still achieving the technical objectives of the system, then you may accept this socio-technical solution as your final solution.

If you have any doubts about the jobs which your chosen solution will create, go back to your short-list of socio-technical solutions, and take the solution which ranks next and proceed with Steps 5 and 6 again, until you are satisfied with your solution.

INTERNATIONAL BANK

DETAILED WORK DESIGN

INTERNATIONAL BANK

DETAILED WORK DESIGN

INTERNATIONAL BANK

STEP 6 PREPARING A DETAILED WORK DESIGN FROM THE CHOSEN SOCIO-TECHNICAL SOLUTION

We have chosen the following socio-technical solution as the most suitable.

T5. Visual display units on-line for input.
Real-time update for deals.
Batch processing for bank information.

S4. Longitudinal groups based on currencies for input, output and non-computer tasks.
Each clerk does all tasks including requisitioning resources and problem-solving.
There is a service group which provides services to all currency groups.

In the diagrammatic form we used earlier, the chosen socio-technical system can be shown as follows:

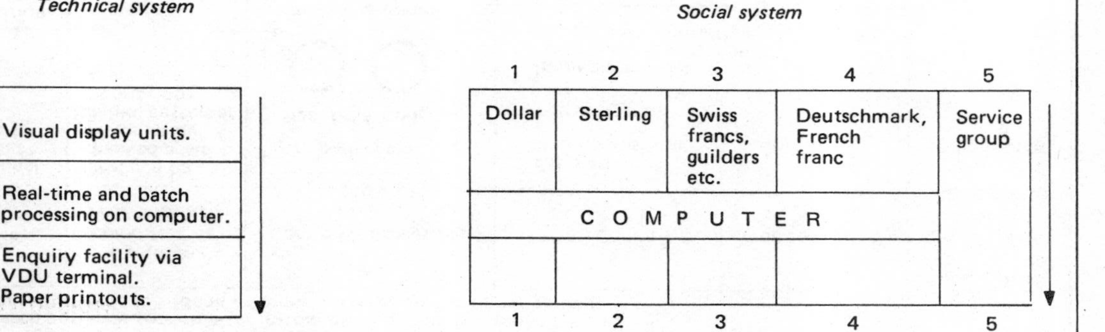

We must now check that this socio-technical solution is the most suitable one to create interesting work for the employees, while still achieving a system which is technically efficient.

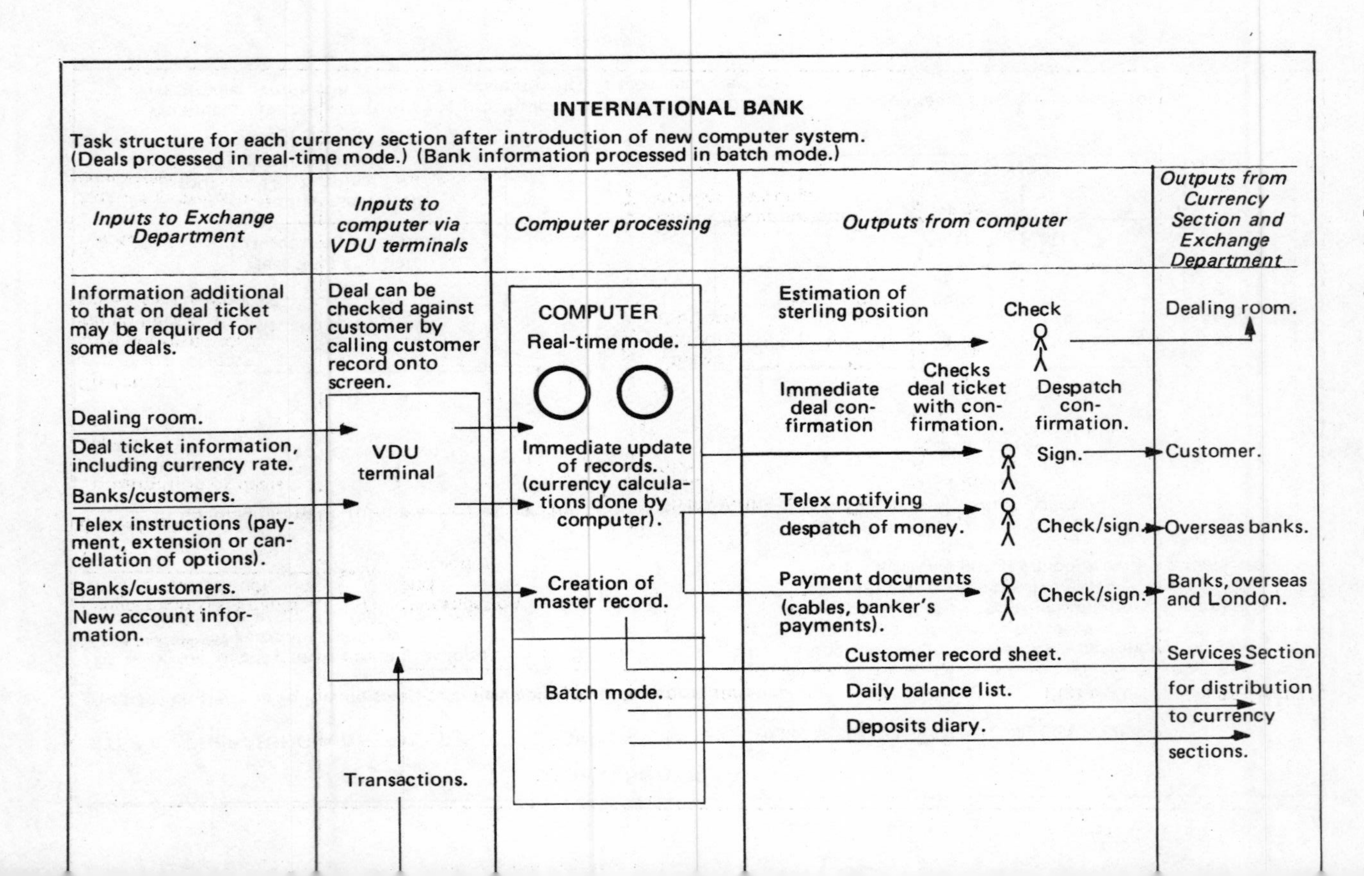
INTERNATIONAL BANK
Task structure for each currency section after introduction of new computer system.
(Deals processed in real-time mode.) (Bank information processed in batch mode.)
Inputs to Exchange Department
Inputs to computer via VDU terminals
Computer processing
Outputs from computer
Outputs from Currency Section and Exchange Department
Information additional to that on deal ticket may be required for some deals.
Deal can be checked against customer by calling customer record onto screen.
COMPUTER
Real-time mode.
Estimation of sterling position
Check
Dealing room.
Immediate deal confirmation
Checks deal ticket with confirmation.
Despatch confirmation.
Dealing room.
Deal ticket information, including currency rate.
VDU terminal
Immediate update of records. (currency calculations done by computer).
Sign.
Customer.
Banks/customers.
Telex instructions (payment, extension or cancellation of options).
Telex notifying despatch of money.
Check/sign.
Overseas banks.
Banks/customers.
New account information.
Creation of master record.
Payment documents (cables, banker's payments).
Check/sign.
Banks, overseas and London.
Customer record sheet.
Services Section for distribution to currency sections.
Batch mode.
Daily balance list.
Deposits diary.
Transactions.

	Processing without computer.	Form.		Services Section
Services Section Ledgers and statements which do not match.	Investigation and correction.			
Other banks. Confirmation of deals handled by these banks.		Check with deals in files.	File.	
Customers. Deal confirmations which have been signed by customers.		Check, correct if effors.	File.	
Banks. Requests for deposit guarantees		Check documentation, correct if wrong.	File.	

INTERNATIONAL BANK

Let us now consider the organisation of work into currency sections in more detail.

1. A study of the number of deals per currency and the time taken to process each type of deal has established that splitting the Exchange Department into four currency sections – dollars, sterling, Swiss francs/guilders/sundry, deutschmark/French francs – and a service section will result in a good work balance between the sections. Each section will require 10 staff, thus saving 18 staff in total. This staff saving is due to the removal of coding and manual checking by the on-line system. The old job titles of coder, checker, etc. will be abandoned and everyone will be called 'currency clerk'. A grading structure will be retained but there will now be only 3 grades below Section Leader. The role of Deputy Section Leader will be abolished.

2. The available mix of tasks comprises:

   (a) Tasks associated with computer input. Collecting information, adding to information, inputting information via VDUs, checking the accuracy of input. When necessary establishing the integrity of input by using the VDU enquiry facility to check on customer status.

   (b) Tasks associated with real-time computer output. Dealing with confirmations and payment documents such as cables and banker's orders. Checking the correctness of these before despatch.

   (c) Tasks associated with batch computer output. Identifying the reason for errors in printed output documents, balance lists etc.

   (d) Tasks still independent of computer. Information coming into the Foreign Exchange Department from other banks, e.g. confirmation of deals handled by other banks, requests for deposit guarantees, etc. Problems associated with meeting customer instructions, reconciling documents, correcting errors.

   (e) Control procedures such as final check and signing before documents are despatched.

INTERNATIONAL BANK

3. These tasks can be available to all clerks in a currency section, the sole criterion for undertaking them being knowledge. The grading and pay structure would, in turn, be related to ability to undertake a growing number of increasingly complex tasks and not to length of service. The youth of the clerical labour force in the Exchange Department makes this approach acceptable. Routine work which cannot be shared amongst the group should be given to part-timers, not young clerks. (The Service Section contains considerable routine work and has less task integration than the Currency Section. This is best staffed by part-time clerks.)

   The objective of this work arrangement is to provide a stimulating learning environment for clerks. One in which they can progressively develop their knowledge.

   Targets for the group will be (a) high quality output with few errors, (b) a multi-skilled membership with a training programme directed at encouraging all clerks to become expert in all jobs. Feedback to the group on how well these targets are being achieved will be the responsibility of the Section Leader.

   The solving of problems will be a group responsibility, in this way providing opportunities for making choices and using judgement.

   The Section Leader's co-ordination responsibility will include seeing that work is divided amongst group members in such a way as to give everyone an interesting, challenging set of tasks. These tasks will be reviewed or reallocated from day to day in the interest both of an efficient use of resources and of training needs.

   The Section Leader has responsibility for ensuring that all group members understand the importance of the set of tasks they are carrying out at any moment in time; and how these tasks fit into the totality of work of the Exchange Department.

4. The Section Leader's role now becomes of very great importance and her responsibilities include maintaining a high morale group, ensuring that the group's training programme is effectively carried out, maintaining a high quality work output and co-ordinating group activity. In view of the importance of controls in banking she must still retain a final checking and signing responsibility.

INTERNATIONAL BANK

The socio-technical solution which has been chosen seems appropriate in that it will enable interesting jobs to be created, at the same time as enabling a technically efficient computer system to be set up. The use of visual display units provides a flexible input medium which assists work accuracy through the immediate identification of errors. The provision of a VDU enquiry facility means that information about customers and deals is immediately available and this will add interest to the job and facilitate the solving of problems.

The longitudinal group arrangements provide a good learning environment and a useful promotion and advancement ladder which should be linked to the grading system.

We can now finally accept this socio-technical solution as the best one possible for undertaking the work of the International Bank.

7 Conclusions: the ETHICS method and its practical applications

We hope that you have enjoyed working your way through the exercises in this book, and that you now have an understanding of one method for re-designing work. Let us now consider how the method was used in these exercises and how it can be used in real change situations.

Objectives of the ETHICS method

One of the main aims of the ETHICS method is to achieve a better balance between technical and social objectives in the design of working systems. As you will have seen, the dual objectives in the exercises were to achieve greater efficiency *and* increased job satisfaction, through careful attention to all aspects of the design process. The problem in achieving well-designed effective systems is not simply that of adjusting people to technology or technology to people. It consists of organising the two so that the best match can be obtained between them. This means that the potential benefit of the technology can be maximised, at the same time as improving the working lives of people using the technology. However, in most change situations, the economic and technical objectives dominate the thinking of those designing the new system. The effects of the system on job design, working relationships and job satisfaction are usually seen as by-products of the system, rather than factors which need to be consciously considered and planned. The specifications for new systems tend to be almost entirely concerned with the technical aspects and very rarely specify the human aspects of the system. Where a system does not have a major impact on people's jobs, this omission may not be too serious. But as computer systems, in particular, become more pervasive in the life of organisations, the effects cannot be left to chance or to *ad hoc* adjustments during implementation. The analysis and specification of the social system and the design of jobs, together with their interrelationships, will become as important as specification of the technical system.

Hopefully, though, a greater concern for the social implications of new systems will not come about merely as a reaction to correct the negative 'side-effects' which often occur. What is needed is a re-evaluation of the currently predominant design philosophy which regards man simply as an extension of the machine. This orientation is particularly strong among systems designers in Britain, where jobs tend to be fragmented into single skills and work is tightly structured and controlled. Such a philosophy has ignored, or failed to see, the fact that new systems provide opportunities actually to improve work design and job satisfaction. When these opportunities are neglected, the full potential benefits of a system may not be attained or indeed, the situation may deteriorate if the match between people and technology becomes worse.

Rather than waste such valuable opportunities, the authors seek to encourage the designers and users of all new systems to deal seriously with the social implications from the very start of the design process. As decisions are gradually taken, the degrees of freedom to modify the system become fewer, until the user may be presented with a *fait accompli,* unless he has been involved throughout the design process. At this stage, some attempt may be made to 'sell' the system to the user, when his active involvement in decision-making would have ensured his commitment to the success of the new system.

By giving greater weight to consideration of the social factors in conjunction with the technical evaluation of system alternatives, the ETHICS method seeks to achieve a wider and more realistic assessment of the implications of new systems. The result of designing systems in this way will be to create jobs which are meaningful and fulfilling. At the same time, such systems are likely to achieve a higher level of technical efficiency than systems which people feel have been imposed on them, and to which they have little personal commitment.

The ETHICS case-studies exercises

Despite the particular relevance of the ETHICS method to the design of new systems, it is important to emphasise that the method is equally applicable to the re-design of existing situations. The desire to achieve a better balance between people and technology may arise because of difficulties which are being experienced in the running of a department, particularly in the dissatisfaction of people working there. In this case, it may not be possible to modify the technical system in any significant way, and an improvement can only be achieved by a careful analysis and re-design of the jobs, to reach a better match with the technology.

The Babycare exercise has shown that it is possible to reorganise work without at the same time making any major change in technology. The stimulus for change in the Babycare situation was an unhappy, low-morale labour force, who complained that the pressure of paced work on the packing line was causing them stress, while at the same time their work was routine and boring. Improvements were achieved in this situation through reorganising the production system, removing machine-pacing and creating a more interesting task structure for the employees.

Inevitably, the redesign of an existing situation is more difficult than designing a new system from the beginning. The constraints created by the technology limit the scope available for improving jobs. Even so, the very fact that significant improvements can be made indicates the degree of flexibility available in situations which initially look most unpromising.

Naturally though, the detailed and thoroughgoing evaluation of technical alternatives and task structures as illustrated in the ETHICS exercise is especially appropriate when a major change in technology is taking place. The exercises presented here have concentrated on computer systems because in recent years there has been such a dramatic increase in the introduction of computer systems which involve complex design procedures. However, the concepts of socio-technical design and achieving the best match between people and technology can apply equally well to the design of new production lines or office procedures which do not involve computers.

However, the authors have most commonly used the ETHICS method when a major change in technology is being made through the introduction of a new computer system. This can either be a firm's first move from a manual to a computer-based system of work, or it may be an upgrading of existing computer technology, for example a move from a batch to an on-line system. In these kinds of situation, the computer acts as a helpful catalyst for achieving social as well as technical change. Most computer systems used in offices alter existing work procedures and if no thought is given to the human impact of the computer, systems can be designed which increase the routinisation of work. Our Mail-It and International Bank exercises have shown that this kind of undesirable consequence can be avoided and that, on the contrary, the introduction of a new computer system can provide an opportunity for acquiring human as well as technical gains.

In the Mail-It case, the introduction of visual display units was carefully planned alongside changes in the organisational structure of the office. The result was that far from increasing the routine, impersonal nature of the work, the new system actually created more interesting jobs with greater responsibility and customer identification. The initial diagnosis of attitudes among the staff contributed a great deal towards recognising the opportunities which were available. It highlighted the desire among staff to move away from simple, repetitive tasks towards a range of tasks through which they could gain a sense of achievement by making a significant contribution towards the work of their department.

The situation in the International Bank exercise was considerably different from Mail-It, in that the work itself was already interesting and complex. The need here was for a simple but efficient computer system which would serve the needs of staff and relieve them of routine tasks. In doing this, it was important that the system should not be too obtrusive, but should rather act as a complement to the specialised knowledge and skills of the Foreign Exchange staff.

The three ETHICS case-study exercises, therefore, present examples of widely differing situations and a range of requirements which the technical systems had to meet. The attitudes of the people involved were also significantly different, though with the shared desire for their needs to be considered at least as important as the technical requirements of the systems. By providing a framework in which the technical system requirements and the job satisfaction needs of people can be evaluated together, the ETHICS method has made a significant contribution towards achieving a better match between them, and so maximising the potential benefits of both.

Groups using the ETHICS method

The authors have used the ETHICS method in a number of different ways, both in the classroom and in practical situations. Broadly speaking, the method has been most helpful for three groups of people, namely, managers, systems designers and system users.

Although managers may not usually need to become so closely involved in the detail of design procedures, it is extremely useful for them to have an appreciation of the concept of socio-technical systems design. Thus when they are faced with a change situation, it is possible for them to have an understanding of the ways in which these methods can be used to develop and evaluate alternative solutions. In contrast with many situations, where technical and economic considerations predominate, the use of the ETHICS method can help managers to see how social factors can also be considered alongside these, in an integrated decision-making process.

For those more directly concerned with the design of new systems, the ETHICS method also provides a useful framework. In particular, we have used the case-study exercises presented here with groups of technical systems designers, to help them think through the social as well as the technical options available. Sometimes this proves quite difficult for them, because they are not used to being asked to consider in this way the needs of people. More often, they tend to regard people as a constraint to be circumvented or as a potential hazard which may interfere with the smooth running of the technical system. Once they have seen the way in which the needs and attitudes of people can be taken into account, in creating a socio-technical system, we have found that systems designers often become very committed to increasing their expertise in understanding social factors. By doing so, they realise that the final result is likely to be more efficient and satisfying than systems designed without these dual objectives.

In general, the managers and systems designers who have used the exercises have done so in the classroom situation. Indeed, the case studies were first developed by the authors for courses at the Manchester Business School, and have been used with many different groups over the years. Our aim was to develop an awareness of the value of the socio-technical approach and to provide a framework for showing how this can be used. Although this can be achieved in the classroom, a much deeper learning takes place when people use the overall framework and apply it to their own particular situation, where they have detailed knowledge and a personal interest. This is especially true for users of the system who are very keen to have the best possible system for their own working situation.

The authors have assisted many firms using participative design approaches, in which groups of clerical or shop-floor employees have played a major role in the design of new work systems. In this kind of situation, the ETHICS method has a number of advantages. Firstly, one of the exercises set out in this book can be used for training purposes, to give employees an understanding of the diagnostic and design processes which they will be carrying out in their own project. Secondly, it provides employees with a set of simple, structured techniques which enable them to tackle without too much difficulty a design activity which is totally new to them. Thirdly, although users may have considerable knowledge of the situation, they may lack some information, especially of human needs, and the job satisfaction diagnosis that forms the first part of the ETHICS approach is a useful way of obtaining this information.

For these user groups, who may have little or no experience of computer systems, the case-study exercises can be a useful introduction, especially if they are guided by a trainer or systems analyst involved in the design process. The case studies can help them to understand that there may be a range of alternative ways of designing the system. Very often, inexperienced users do not realise the possibilities which are available and the different ways in which their work could be organised and are therefore reluctant to voice their opinions. We have found that the use of one of the case-study exercises can be useful to help them recognise how new systems may be designed to help increase their job satisfaction, as well as their effectiveness at work. To assist these less experienced groups, Appendix B includes a very simple outline of the main concepts and terms used to describe computer systems.

Case studies as a learning aid

The development of the case-study method of teaching has been a major contribution of the business schools. It is based on the realisation that learning is more effective when people actually become involved in working through ideas and data themselves, rather than merely being 'taught' by the lecturer. However, there is also a need for this type of case-study material to be made available to a wider audience.

The format of this book has been specially developed so that individuals and groups who are interested in the concepts of socio-technical systems design can work through a variety of cases themselves. For individuals, this means that they can proceed at their own pace, and perhaps use these exercises in conjunction with some of the 'teach-yourself' cassette courses which are available.

In general, though, the group-learning situation can provide a wider range of experience and ideas. People tend to learn more quickly when they share and discuss the exercises together, and we would encourage trainers to use the case studies on a group basis whenever possible.

Appendix A gives an outline of the way the ETHICS case-study exercises may be used with a group of students, including a suggested time-table and a note of the facilities and equipment which should be ideally provided.

Socio-technical design in the real world

It must be stressed that the approach is never quite as simple and clear-cut in real world situations as it appears in our exercises. To some extent, all case studies have to present a straightforward, even rather artificial, view in order to highlight the points which the case study is intended to illustrate.

In reality, socio-technical systems design is much more of an iterative process than we have shown here, particularly when technical and social options are being generated. We have formulated the case studies in terms of a series of social and technical 'solutions', to explain the step-by-step procedures of the ETHICS method, but in fact, real life situations are far more complex than this suggests. Each 'solution' may include a whole range of options, which cannot be realistically worked through in a case-study exercise, but may be extremely important in the evaluation of a particular system.

The process of investigating alternatives and working through the technical and social advantages of these tends to suggest other alternatives, which then have to be considered.

Also, in the exercises, we have set out the social and technical advantages and disadvantages as if these are always known. In fact, evaluating a new piece of technical equipment may take many months and it may not always be possible to understand and predict precisely what its technical and social advantages will be. Similarly, as we have found when helping technical experts to use our method, a socio-technical design which objectively appears to be excellent may be rejected by users because they would prefer something different.

This is a strong argument for using a participative approach to systems design, so that users or their representatives can do their own analysis and arrive at their own preferred solution. However, many systems analysts and managers may feel that such a participative involvement of users at the design stage may increase the planning time and costs of the system. Although this may be the case, our experience is that systems designed without the active involvement of users may initially appear to be cost-effective on technical criteria, but in fact often incur high social costs, such as resistance to change, poor equipment utilisation, high turnover and absenteeism. The total cost of such a system could well be considerably more than if users had themselves helped to design the system which fitted their job-satisfaction needs and to which they felt a sense of personal commitment.

In developing the ETHICS method and preparing the three illustrative case studies presented here, we have sought to show that the integration of social and technical objectives in the design of systems is feasible and indeed desirable. Unless specific social objectives are established alongside the technical objectives for a new system, the social factors are likely to be largely ignored or treated as an afterthought. But managers, systems analysts and user departments are gradually becoming more aware of the possibility of creating effective socio-technical systems, to increase both job satisfaction and technical efficiency. We hope that this book may contribute towards developing that awareness and in creating a new, more enlightened approach to systems design.

Appendix A

Notes for using the exercises with groups of students

The best size of group is from four to eight people and if a number of groups take part then the exercise can be made a competitive one.

For each exercise first photocopy all pages except those with suggested answers and assemble these as a set of instructions and blank work sheets. When students have completed the exercise and presented and discussed their solutions, photocopy and hand out pages with suggested answers for insertion into the exercise. If required, copies of the work sheets and the solutions can be obtained from Enid Mumford or Mary Weir.

Each group of students should be provided with a blackboard, and if possible, some transparencies for use with an overhead projector. Each member of each group should have a copy of the instructions and exercise documents, plus plenty of scrap paper.

To do a full case study properly, without feeling pressured, usually requires a day's work, although an experienced group can work more quickly, and we have occasionally managed to complete a case study in half a day. However, if time is restricted we find it is preferable to ask students to do only part of the case study, and to make sure they do it in detail. It is also essential to allow adequate time for a feedback session and this may suffer if students are given too much work to do in the time available.

Time-table for running a case study

The groups with whom we have run the case studies have often had management and/or systems analysis experience, and so they are familiar with the concepts of organisational analysis, and can apply these to the case studies. Where students have not had this experience, and increasingly we find we are using the approach to give clerical and shop-floor groups an understanding of socio-technical design, the tutor may feel that they need some preliminary teaching sessions before they tackle the case study, especially on the ideas associated with socio-technical systems design and motivation theory.

If students are to be allowed less than a day for the case study, it is an advantage if they can be given the documents some time beforehand with a very brief explanation. This gives them a chance to read the documents through, and they attend to do the case study already well prepared. Where a full day has been set aside for the exercise, then this preparatory reading is not essential.

It is an advantage to start the day with a short formal talk, perhaps only fifteen minutes, reminding students of the general theoretical background which they will be using during the case study exercise. This should be followed by a description of the ETHICS method as explained in Chapter 3 of this book. Students often find it helpful if they have the instruction sheets in front of them at this stage so they can see how the case study fits into the method as a whole.

The tutor should then go on to explain what the students will be doing during the case-study exercise, by going through the instructions step by step, and linking this explanation with the documents which students have been given. At this stage, some background information about the industry and type of system involved in the case-study is very valuable, though not essential, and of course the tutor can pitch this information at a suitable level for his students.

All these preliminary stages need not take more than 30-45 minutes to complete, before the students are asked to form the small groups in the rooms which have been allocated to them. Before they depart they should be told approximately how long they should take for the different stages of the case-study, and they should be asked to appoint a spokesman who will report back at the feedback session. The tutor should also assure them that he will make frequent visits to the groups to give any assistance which may be necessary.

It is very helpful if the tutor can visit each group fairly soon after they begin the case study, to clear up any problems quickly and make sure they get off to a good start. We often find that if groups do not make much headway at first, they become despondent and have difficulty in catching up later. As a rough guide, groups appreciate a visit every 30-60 minutes depending on their level of competence. If they are only doing part of a case study and taking only a short time over it, then they are likely to need more frequent visits to make sure they get through in the time allowed.

When the groups have completed the case study, or reached the time set for the feedback session, they should be asked to reassemble together. The spokesman of each group should then be asked to present his group's solution, usually by making use of the blackboard, or by showing transparencies which the group have prepared, on an overhead projector. The latter method saves time, but often does not lead to such clear expositions as the blackboard method because the answer tends to be presented all at once rather than in logical stages.

It is best to allow each group about 15-20 minutes to present their solutions. We usually let all the groups present their solutions before making comments or starting discussion, apart from asking for clarification of any points which seem to require it. If the tutor takes the feedback in this way, then he will have to make notes during the presentation and also write down his comments so he can sum up at the end. Alternatively he could comment after each presentation, while the answers are still on the blackboard. This is considerably easier for the tutor, but does not allow him to make comparisons between the answers. If the provision of blackboard space is generous, the best of all worlds is to leave the answers for all the groups on the blackboard. This makes it much easier to comment at the end, but it is not always possible especially if there is only one fairly small blackboard available.

Once all the presentations have been made, the tutor can comment on the answers which have been given, draw out the points which he thinks are important, and clarify any difficulties which have arisen. He may wish to open a discussion between the students on the exercise and the ideas behind it, if it seems that they will benefit and time permits. He can then go on to ask for examples from the students' own experience where they could use these ideas to advantage.

The tutor can hand out a suggested solution either before the feedback session, before he comments on the presentations, or afterwards, as a summing up. He may wish to use this in the discussion, as a solution which all groups can comment on, without feeling that any of them is personally involved with it.

The follow-up after the case study will, of course, depend on individual circumstances and should be decided by the tutor. The students may benefit from doing the other case studies presented here, or from preparing examples from their own situations. In some cases further lectures may be useful, to go into greater depth, once the basic ideas have been conveyed to the students.

The time-table for a typical day's case-study exercise with four groups might be as follows:

9.00 - 9.45	Introduction by tutor.
9.50 - 10.30	Groups.
10.30 - 10.45	Break.
10.45 - 1.00	Groups.
1.00 - 2.00	Lunch.
2.00 - 3.30	Groups.
3.30 - 3.45	Break.
3.45 - 5.00	Feedback and discussion.
5.00	Final summing up by tutor.

This time-table allows the group a total of about 4½ hours to work together on the case-study, and 1¼ hours for feedback and discussion. Of course such a time-table could be adjusted to suit the particular needs of students and circumstances in which the exercise is being used.

The competence level of groups doing the case study

As mentioned earlier, we have run these case-studies with groups from very different backgrounds, and at various levels of competence in the areas considered here. In order to allow all the groups to undertake work within their abilities, we tailor a case study specifically for each group, from the basic material presented here. In fact, in situations where we were teaching a group with whom we were not very familiar, and therefore could not gauge their competence in doing the case study, we have often taken along two versions. We started groups on the more difficult version and if they could cope with it, they were allowed to proceed. However, if they complained of difficulty, or were working rather too slowly for the time allowed, we handed them a simpler version to complete and this usually solved any problems they may have been experiencing.

We anticipate that any tutor using the case-study exercises would need to do the same tailoring operation, to make it applicable to the needs of his particular students. The two main things he would need to take into account are the time available for the exercise and the abilities and experience of his students.

There are several possible ways of arranging the material in the exercise and it is up to the tutor to select the combination most appropriate for his group. Depending on the time available for the exercise, one of the three stages can be selected, as follows:

1. Students can do the whole case study.
2. Only the *social* solutions need be worked out, with a description of the technical system being given. This is really an exercise in what we have called systems adjustment.

3. The work design step on its own provides a useful short exercise on this topic.

There are also several ways in which the tutor could make the exercise more or less difficult, depending on the abilities of his students, for example:

1. The most difficult exercise is to ask students to generate all technical and social solutions themselves, and work out their advantages and disadvantages. This would suit a very competent group.
2. People with a technical background could be given the technical solutions and evaluations which would be easy for them, and asked to complete only the social solutions.
3. An alternative way is to give students the whole case study, but to make it easier, is to provide a description of all the social and technical solutions, and ask them to decide on the advantages and disadvantages of each and then complete the exercise.
4. Groups which are very new to the concepts being used here, or want to go through the whole case study but are very short of time, could be given all the technical and social solutions with their advantages and disadvantages and asked to select from them the most appropriate socio-technical solution.

Appendix B

Glossary of computer terms used in the case studies

The purpose of this appendix is to provide a very simple outline of the main concepts and terms used to describe computer systems, for those people who, as yet, have had little contact with this technology. Since the main focus of the case studies is to improve the design of work done by users of computer systems, the terms included here will naturally concentrate on the input and output devices used, and the modes of operation, as these have the most important implications for the jobs of users, rather than the actual operation of the computer itself.

The basic features of computer systems

All computer systems have the same basic features, as follows:

Input units. To read data and programs into the computer.

Store (memory). Stores data, instructions and immediate results.

Arithmetic unit. To perform calculations and process data.

Control unit. To interpret instructions and implement them by commands in the form of an electrical pulse.

Output units. To present the results of a computer program in a machine-readable or human-readable form.

The input and output units are referred to as the 'peripherals' of the computer, and the control unit, arithmetic unit and store are together called the central processing unit or CPU. These basic features are presented diagrammatically in Figure A.1. As far as the users in our case studies are concerned, they are mainly affected by:

1. The selection of different types of input device.
2. The selection of different types of output device.
3. The mode of operation of the computer itself.

These will now be dealt with in turn.

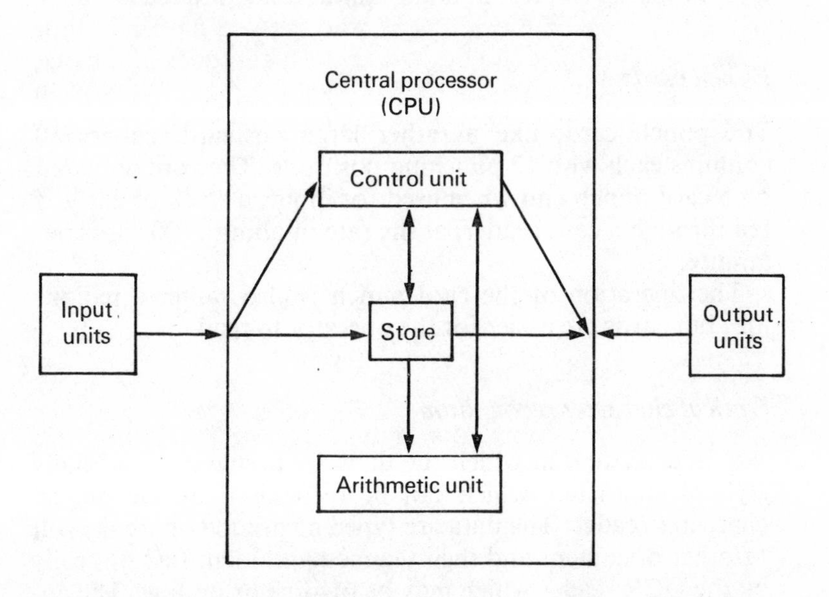

Figure A.1 Basic features of computer systems

Input devices

Paper tape punch

Paper tape is a narrow ribbon of paper carrying punched holes representing data, similar to that used in the Telex system. The data are recorded more or less continuously along the length of the tape, each character being represented by a particular pattern of holes across the width of the tape. The tape is prepared by key punching on a paper tape punch and is verified on a special verifying machine. When used as input, the tape is fed into a tape-reader which optically senses the presence or absence of holes.

The operation of the key punch can be extremely repetitious and boring and the paper tape which is produced is not a very exciting end result, and is quite difficult to read.

Punch cards

The punch card, like a rather large postcard, carries 80 columns each with 12 punching positions. They are prepared on a card punch and when used for input, a stack of cards is fed through a card reader, at the rate of about 1200 cards per minute.

The operation of the card punch is also rather a tedious job, but cards are easier for the operator to read.

Optical character recognition

An input system in which the data are prepared as specially stylised characters which can be recognised by the optical character reader. The data are typed or printed on a tally roll or other document and then scanned and identified optically by the OCR reader which may be off-line or on-line. The advantage of this system for the user is that the data can be prepared in a human-readable form as well as being machine-readable. However, the reading process is rather slow and tolerances are tight, so that the system tends to be expensive.

A variant of this method is optical mark sensing where a

mark is made in pencil in a predetermined position on a specially prepared form, such as an order form.

Key to tape

A development of the key punch, where data are typed *directly* on to magnetic tape rather than on to paper tape or punched cards. This system is faster and more economical to operate, since it cuts out one stage of the old data preparation system.

Key to disk

In this system, the data are keyed directly on to a small disk, again cutting out an intermediate stage. The data are keyed in and are displayed on a small screen for verification. Once verified, the data are transferred to the magnetic disk. By using a small computer for verification and editing data in this way, the process of data preparation is much faster and simpler. The operator can interact directly with the computer in preparing error-free data and can visually check, since the data are presented in a human-readable form.

Visual display units (VDUs)

A computer terminal incorporating a video display for showing diagrammatic or alphanumeric information. They are mainly used for interactive systems where the output may not be needed in a permanent form. Where a permanent copy is required, a typewriter terminal may be more appropriate, although a 'hard copy' printer may be attached to a VDU.

The VDU has largely taken the place of the typewriter terminal because it is faster, quieter and pleasanter to use, since information is displayed quickly in a form which is easily understandable to the user.

Output devices

There is a wide range of equipment by which the computer transmits information to the outside world:

Line printer output

The majority of output is produced as 'hard-copy' or paper printout on to continuous stationery. The line printer, as its name implies, prints a whole line of information at a time and can operate at a very high speed, still producing high-quality printing.

Other types of hard-copy printer are the page-printer, drum printer, chain printer, belt printer. Non-impact printers use xerography rather than continual mechanical methods of printing, and can achieve much higher speeds.

The problem with this type of output for some applications, is the volume of paper which is produced, requiring to be filed and stored, and which can be heavy to handle.

Computer output microfilm (COM)

A system of producing computer output directly on to microfilm. This can then be read directly through a viewer. Although there has been some reluctance on the part of users to accept this system, microfilm is far more compact to handle and store than the more usual paper printout.

Terminals

Increasingly, data are being output via terminals directly to the users, whether via hard-copy typewriter terminals, or through visual display units, where a permanent record is not required. As more systems operate on-line and even on-line real-time, greater use is being made of remote terminals, that is, located at the remote end of a data transmission link, often far away from the computer itself.

Many terminals are known as 'intelligent terminals' which

means they often incorporate a microcomputer which enables them to perform sort and edit procedures. Specialised terminals have been developed for airline booking systems, banking, retail point-of-sale systems, etc.

Modes of operation of the computer

It is important to understand not only the methods by which data is transferred to and from the computer, but also the main 'modes' of operation by which this is done. Very broadly, there are three main stages involved in data processing.

(a) Data preparation.
(b) Updating permanent computer files.
(c) Output of new data.

Data preparation

Off-line editing. Before data from the source document are read into the computer, they have to be edited to eliminate errors. In the simple paper tape or punched card systems, this is done by a preliminary editing computer run, which produces an error report. The data have then to be repunched and verified before being re-input. Clearly, therefore, it can be a time-consuming exercise to do data preparation in this 'off-line' mode.

On-line editing. It is now more usual to have an 'on-line' editing system. This means that the data are checked as they are being input so errors can be dealt with immediately before going on to the magnetic tape or disk. This interactive method of data preparation eliminates the special verification stage and the editing run of the data. 'Key to tape' and 'key to disk' methods of input are used for this type of on-line editing.

Updating the data

Batch processing. The tape or disk of edited data is then subsequently used to update the main computer files, when a batch of correct data has been input. The updating of files in this way is called 'batch processing' because it is only done at intervals when a batch of data is ready for processing.

On-line real-time. Still quicker and simpler is to combine the editing and updating into one operation in an on-line system, where the terminal has direct access to the main computer and data can be checked against the central files to see if they are correct before being used immediately to update those files. This mode of operation is referred to as 'on-line real-time', since the data editing, updating and output are done so rapidly that the user perceives them as one operation, and has immediate output.

Output of new data

A major difference in these systems for the user is the 'response time', that is the speed at which he gets the updated information from the computer. In an 'off-line' system, the response may be 24 hours or sometimes longer. An 'on-line' system can give a rapid response about the current position, but may not be able to deal with a new transaction because the files are not updated until some time later. On the other hand, an 'on-line real-time' system can also provide information on the new position immediately after a new transaction has taken place, because the user has access to the main computer file for updating. In this case, the response time is measured in seconds, not hours. Air-line reservation systems, banking systems and process control systems are examples of situations where a real-time response is required to provide the best possible service from the computer.

Adapting systems to the user

As far as the user is concerned, the systems which can read human-readable input and output and provide a fast response time are generally pleasanter to operate. They are better suited to human capabilities and mental capacities than systems which require input which is difficult for people to read, and where the response is so slow that the immediate interest of the user is lost by the time the updated information is received. Of course, on-line and on-line real-time systems are much more complex and expensive to set up and operate and it is here that the priorities of the systems designer are important, in balancing the needs of the social and technical systems in achieving a cost-effective total socio-technical computer system.

References

1. Argyris, C., *Intervention Theory and Method,* Addison-Wesley (New York 1970).
2. Blake, R.N. and Mouton, J.S., *The Management Grid,* Gulf (Houston 1964).
3. Blauner, R., *Alienation and Freedom,* University of Chicago Press (Chicago 1964).
4. Bowey, A. and Lupton, T., 'Productivity drift and the structure of the pay packet', *Journal of Management Studies,* Vol. 7, No. 2, May 1971.
5. Bowey, A., 'Tailor-made payment systems and employee motivation', *Management Decision,* Vol. 10, No. 1, 1972.
6. Cooper, R. and Foster, M., 'Socio-technical systems', *American Psychologist,* Vol. 26, No. 5, 1971.
7. Cooper, R., 'How jobs motivate', *Personnel Review,* Vol. 2, No. 2, 1973.
8. Crozier, M., *The Bureaucratic Phenomenon,* Tavistock (London 1964).
9. Davis, L.E., *The Design of Jobs,* Penguin (Harmondsworth 1972).
10. Davis, L.E., 'Job satisfaction research: the post-industtrial view', *Industrial Relations,* Vol. 10, 1971.
11. Edwards, G. 'Manufacturing — not so much a technology, more a way of life', *Personnel Review,* Vol. 2, No. 2, 1973.
12. Gouldner, A.W., *Patterns of Industrial Bureaucracy,* Routledge and Kegan Paul (London 1955).

13. Gowler, D. and Legge, K., 'The wage payment system: a primary infrastructure', in *Local Labour Markets and Wage Structures,* Gower (London 1970).
14. Gulowson, J. 'Organisation design in industry — towards a democratic, socio-technical approach', *Personnel Review,* Vol. 2, No. 2, 1973.
15. Herbst, P.G., *Socio-technical Design,* Tavistock (London 1972).
16. Herzberg, F., *Work and the Nature of Man,* Staples (London 1966).
17. Likert, R., *The Human Organisation,* McGraw-Hill (New York 1967).
18. Lupton, T. and Gowler, D., *Selecting a Wage Payment System,* Federation research paper 3, Engineering Employers (London 1969).
19. Martin, J. and Norman, R.D., *The Computerised Society,* Prentice-Hall (Englewood Cliffs, NJ 1970).
20. Maslow, A.H., *Motivation and Personality,* Harper (New York 1954).
21. McGregor, D., *The Human Side of Enterprise,* McGraw-Hill (New York 1960).
22. Meister, D. and Radieau, G.F., *Human Factors Evaluation in System Development,* Wiley (New York 1965).
23. Mumford, E., 'A new approach using an old theory', *Sociological Review,* Vol. 18, No. 1, March 1970.
24. Mumford, E., *Systems Design for People,* National Computing Centre (Manchester 1971).
25. Mumford, E., Mercer, D., Mills, S. and Weir, M., 'The Human Problems of Computer Introduction', *Management Decision,* Vol. 10, No. 1, 1972.
26. Mumford, E., *Job Satisfaction: A Study of Computer Specialists,* Longmans (London 1972).
27. Mumford, L., *The Pentagon of Power,* Harcourt Brace, Jovanovich (New York 1970).
28. Parson, T. and Shils, E. (eds.) *Towards a General Theory of Action,* Harvard University Press (Cambbridge, Mass. 1951).
29. Paul, W.J. and Robertson, K.B., *Job Enrichment and Employee Motivation,* Gower (London 1970).

30. Sackman, H., *Computers, System Science and Evolving Society,* Wiley, (New York 1967).
31. Sayles, L.R., *Behaviour of Industrial Work Groups,* Wiley (New York 1958).
32. Thorsrud, E., 'Socio-technical approach to job design and organisational development', *Work Research Institute,* Oslo, 1967.
33. Trist, E.L., Higgins, E.W., Murray, H. and Pollock, A.B., Organisational Choice, *Tavistock,* (London 1963).
34. Woodward, J., *Industrial Organisation,* Oxford University Press, (Oxford 1965).

Index

Alienation, 8
Autonomous work groups, 23, 32

Blake, R.N., 12
Bowey, A., 12, 13
Bureaucracy, 12

Cherns, A., 13
Cooper, R., 13, 22
Crozier, M., 12, 13

Davis, L., 6, 13
Design philosophy, 2, 3, 7, 288
Diagnosis, 26, 35, 38

Efficiency needs, 19, 38
Engineering systems, 8
Engineers, 9, 16

Gouldner, A., 13
Gowler, D., 12, 13
Group technology, 8
Gulowsen, J., 13

Herbst, P., 13
Herzberg, F., 11, 18, 19
Human alternatives, 28, 30, 40
Human needs, 3, 10
Humanistic automation, 10

Industrial relations, 9

Job design (see Work design)
Job enlargement, 3, 30
Job enrichment, 3, 8, 30
Job expectations, 11, 15, 16, 25
Job experience, 15, 16
Job requirements, 11, 16, 27
Job rotation, 3, 30

Job satisfaction, 8, 10, 11, 12, 13, 15, 16, 21, 25, 26, 27, 34, 40, 287

Knowledge needs, 18, 38

Legge, K., 12, 13
Likert, R., 11
Lupton, T., 12, 13

Man-machine systems, 7
Managers, 3, 9, 16, 291, 295
Manchester Business School, 4, 13, 292
Maslow, A., 11, 19
Motivation, 11, 12
Moulton, J.S., 12
Mumford, L., 5

Objectives, 3, 5, 6, 9, 10, 29, 33, 39, 287

Parsons, Talcott, 13, 14
Participation, 8, 24, 292, 295
Pay systems, 20
Psychological needs, 19, 38

Questionnaires, 27

Sackman, H., 10
Social science, 2
Social scientists, 8
Social systems, 6
Socio-technical approach, 6, 18, 22, 41
Socio-technical systems, 28, 29, 35, 39, 43, 291, 292, 294
Systems design, 6, 7, 26, 33, 293
Systems designers, 7, 9, 16, 288, 291, 295
Systems evaluation, 26, 34, 288
Systems monitoring, 26, 33

Task structure, 21, 22, 38, 290
Tavistock Institute, 13
Technical alternatives, 28, 30, 40, 290
Technical needs, 3, 10
Technology, 3, 6, 8, 10, 21, 287
Thorsrud, E., 13
Trade unions, 9

Values, 6, 9, 12, 16, 23, 24

Wilson, T., 13
Work controls, 20
Work design, 2, 5, 6, 9, 13, 22, 41, 287
Work organisation, 8
Work Research Unit, 13